# GUIDED NOTEBOOK

## GEORGE WOODBURY

# INTERACTIVE STATISTICS
## SECOND EDITION

## Michael Sullivan, III
*Joliet Junior College*

## George Woodbury
*College of the Sequoias*

 Pearson

The author and publisher of this book have used their best efforts in preparing this book. These efforts include the development, research, and testing of the theories and programs to determine their effectiveness. The author and publisher make no warranty of any kind, expressed or implied, with regard to these programs or the documentation contained in this book. The author and publisher shall not be liable in any event for incidental or consequential damages in connection with, or arising out of, the furnishing, performance, or use of these programs.

Reproduced by Pearson from electronic files supplied by the author.

ISBN-13: 978-0-13-472240-5
ISBN-10: 0-13-472240-X

# Contents

# Chapter 1 – Data Collection

## OUTLINE

**1.1** Introduction to the Practice of Statistics
**1.2** Observational Studies versus Designed Experiments
**1.3** Simple Random Sampling
**1.4** Other Effective Sampling Methods
**1.5** Bias in Sampling
**1.6** The Design of Experiments

## Putting It Together

Statistics plays a major role in many aspects of our lives. It is used in sports, for example, to help a general manager decide which player might be the best fit for a team. It is used in politics to help candidates understand how the public feels about various policies. And statistics is used in medicine to help determine the effectiveness of new drugs.

Used appropriately, statistics can enhance our understanding of the world. Used inappropriately, it can lend support to inaccurate beliefs. Understanding statistical methods will provide you with the ability to analyze and critique studies and the opportunity to become an informed consumer of information.

Understanding statistical methods will also enable you to distinguish solid analysis from bogus "facts."

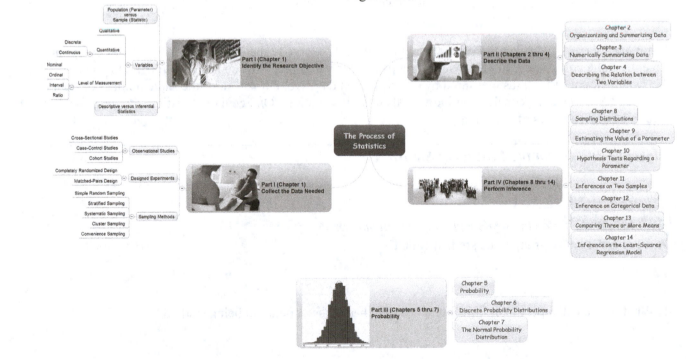

## Section 1.1
## Introduction to the Practice of Statistics

**Objectives**

❶ Define Statistics and Statistical Thinking

❷ Explain the Process of Statistics

❸ Distinguish between Qualitative and Quantitative Variables

❹ Distinguish between Discrete and Continuous Variables

❺ Determine the Level of Measurement of a Variable

### *Objective 1: Define Statistics and Statistical Thinking*

OBJECTIVE 1, PAGE 1

*Answer the following as you watch the video.*

1) Write the definition of statistics below.

2) Data describes _____ of individuals and can be either _____ or _____.

**Note:** Data varies. Consider the students in your class. Is everyone the same height? No. Does everyone have the same color hair? No. So, within groups there is variation. Now consider yourself. Do you eat the same amount of food (as measured by calories) each day? No. Do you sleep the same number of hours each day? No. So, even considering individuals there is variation. One goal of statistics is to describe and understand sources of variation.

### *Objective 2: Explain the Process of Statistics*

OBJECTIVE 2, PAGE 1

*Answer the following while watching the animation.*

3) What is the entire group to be studied called?

4) What do we call a person or object that is a member of the population being studied?

OBJECTIVE 2, PAGE 1 (CONTINUED)

5) Give the definition of a sample.

6) What do we call a numerical summary of a sample?

7) What do we call a numerical summary of a population?

8) Give the definition of descriptive statistics.

9) Give the definition of inferential statistics.

10) In the $100 experiment, what is the population? What is the sample?

Population:

Sample:

11) Is the statement an example of descriptive statistics or inferential statistics? Circle the correct answer.

    A) The percent of students in the survey who would return the money to the owner is 78%.

        Descriptive statistics              Inferential statistics

    B) We are 95% confident that between 74% and 82% of all students would return the money.

        Descriptive statistics              Inferential statistics

12) Is the given measure a statistic or a parameter? Circle the correct answer.

    A) The percentage of all students on your campus who own a car is 48.2%.

        Statistic                    Parameter

    B) Suppose a random sample of 100 students is obtained, and from this sample we find that 46% own a car.

        Statistic                    Parameter

OBJECTIVE 2, PAGE 7

*Fill in the following steps while watching the video.*
**The Process of Statistics**

1. _____
A researcher must determine the question(s) he or she wants answered. The question(s) must be detailed so that it identifies the population that is to be studied.

2. _____
Conducting research on an entire population is often difficult and expensive, so we typically look at a sample. This step is vital to the statistical process because if the data are not collected correctly, the conclusions drawn are meaningless. Do not overlook the importance of appropriate data collection.

3. _____
Descriptive statistics allow the researcher to obtain an overview of the data and can help determine the type of statistical methods the researcher should use.

4. _____
Apply the appropriate techniques to extend the results obtained from the sample to the population and report a level of reliability of the results.

**Example 1**     *The Process of Statistics: Gun Ownership*

The AP – National Constitution Center conducted a national poll to learn how adult Americans feel existing gun-control laws infringe on the second amendment to the U.S. Constitution.
The following statistical process allowed the researchers to conduct their study.

1. Identify the research objective.

2. Collect the information needed to answer the question posed in (1).

3. Describe the data.

4. Perform inference.

---

*Objective 3: Distinguish between Qualitative and Quantitative Variables*

Define the following terms.

13) Qualitative variable:

14) Quantitative variable:

OBJECTIVE 3, PAGE 2

---

**Example 2**     *Distinguishing between Qualitative and Quantitative Variables*

Determine whether the following variables are qualitative or quantitative.

A) Gender

B) Temperature

C) Number of days during the past week that a college student studied

D) Zip code

---

## Objective 4: Distinguish between Discrete and Continuous Variables

OBJECTIVE 4, PAGE 1
Define the following terms.

15) Discrete variable:

16) Continuous variable:

<u>OBJECTIVE 4, PAGE 2</u>

**Example 3**     *Distinguishing between Discrete and Continuous Variables*

Determine whether the quantitative variables are discrete or continuous.

A) The number of heads obtained after flipping a coin five times.

B) The number of cars that arrive at a McDonald's drive-through between 12:00 P.M. and 1:00 P.M.

C) The distance a 2011 Toyota Prius can travel in city driving conditions with a full tank of gas.

<u>OBJECTIVE 4, PAGE 4</u>
Define the following terms.

17) Data:

18) Qualitative data:

19) Quantitative data:

20) Discrete data:

21) Continuous data:

**Example 4**     *Distinguishing between Variables and Data*

The following table presents a group of selected countries and information regarding these countries as of September, 2010.

| Country | Government Type | Life Expectancy (years) | Population (in millions) |
|---|---|---|---|
| Australia | Federal parliamentary democracy | 81.63 | 21.3 |
| Canada | Constitutional monarchy | 81.23 | 33.5 |
| France | Republic | 80.98 | 64.4 |
| Morocco | Constitutional monarchy | 75.47 | 31.3 |
| Poland | Republic | 75.63 | 38.5 |
| Sri Lanka | Republic | 75.14 | 21.3 |
| United States | Federal republic | 78.11 | 307.2 |

Identify the individuals, variables, and data.

---

## *Objective 5: Determine the Level of Measurement of a Variable*

List the characteristics used to determine what level of measurement a variable is.

22) Nominal:

23) Ordinal:

OBJECTIVE 5, PAGE 1 (CONTINUED)
24) Interval:

25) Ratio:

OBJECTIVE 5, PAGE 2

**Example 5**    *Determining the Level of Measurement of a Variable*

For each of the following variables, determine the level of measurement.
A) Gender

B) Temperature

C) Number of days during the past week that a college student studied

D) Letter grade earned in your statistics class

## Section 1.2
## Observational Studies versus Designed Experiments

**Objectives**

❶ Distinguish between an Observational Study and a Designed Experiment

❷ Explain the Various Types of Observational Studies

---

### Objective 1: Distinguish between an Observational Study and a Designed Experiment

OBJECTIVE 1, PAGE 1

*Answer the following as you watch the video.*

1) Why is the Danish study mentioned in the video an observational study and not a designed experiment?

2) Why is the "rat" study mentioned in the video a designed experiment and not an observational study?

3) What is the response variable in each study, and what is the explanatory variable?

OBJECTIVE 1, PAGE 2

*Answer the following after watching the video.*

4) In research, we wish to determine how varying an explanatory variable affects …

5) What does an observational study measure? Does an observational study attempt to influence the value of the response variable or explanatory variable?

OBJECTIVE 1, PAGE 2 (CONTINUED)

6) Explain how you would determine if a study is a designed experiment.

OBJECTIVE 1, PAGE 4

*Watch the video and answer the following.*

7) Why is the influenza study mentioned in the video an observational study and not a designed experiment?

8) List some changes that could be made to investigate the effectiveness of the flu shot with a designed experiment.

9) List some lurking variables in the influenza study.

10) What are some variables (besides getting a flu vaccine) that may play a role in whether one contracts pneumonia or influenza?

OBJECTIVE 1, PAGE 4 (CONTINUED)
11) Define confounding in a study.

12) What is a lurking variable?

13) Do observational studies allow a researcher to claim causality?

OBJECTIVE 1, PAGE 7
14) List some reasons why an observational study would be conducted if causation cannot be claimed.

OBJECTIVE 1, PAGE 8
15) Define: Confounding variable

OBJECTIVE 1, PAGE 9
16) What is the big difference between lurking variables and confounding variables?

## *Objective 2: Explain the Various Types of Observational Studies*

<u>OBJECTIVE 2, PAGE 1</u>

*Answer the following while watching the video.*

17) Define: Cross-sectional studies

18) Define: Case-control studies

19) List some difficulties that may occur and affect the outcomes of a case-control study.

20) List some of the advantages of performing a case-control study over a cross-sectional study.

21) Define: Cohort studies

22) List an advantage of using a cohort study.

23) List two disadvantages of using a cohort study.

| Example 1 | *What Type of Study?* |

Determine whether each of the following studies depict an observational study or an experiment. If the researchers conducted an observational study, determine the type of the observational study.

A) Researchers wanted to assess the long-term psychological effects of children evacuated during World War II. They obtained a sample of 169 former evacuees and a control group of 43 people who were children during the war but were not evacuated. The subjects' mental states were evaluated using questionnaires. It was determined that the psychological well being of the individuals was adversely affected by evacuation. (Source: Foster D, Davies S, and Steele H (2003) The evacuation of British children during World War II: a preliminary investigation into the long-term psychological effects. Aging & Mental Health (7)5.)

B) Xylitol has proven effective in preventing dental carries (cavities) when included in food or gum. A total of 75 Peruvian children were given milk with and without xylitol and were asked to evaluate the taste of each. Overall, the children preferred the milk flavored with xylitol. (Source: Castillo JL, et al (2005) Children's acceptance of milk with xylitol or sorbitol for dental carries prevention. BMC Oral Health (5)6.)

C) A total of 974 homeless women in the Los Angeles area were surveyed to determine their level of satisfaction with the healthcare provided by shelter clinics versus the healthcare provided by government clinics. (Source: Swanson KA, Andersen R, Gelberg L (2003) Patient satisfaction for homeless women. Journal of Women's Health (12)7.)

D) The Cancer Prevention Study II (CPS-II) is funded and conducted by the American Cancer Society. Its goal is to examine the relationship among environmental and lifestyle factors on cancer cases by tracking approximately 1.2 million men and women. Study participants completed an initial study questionnaire in 1982 providing information on a range of lifestyle factors such as diet, alcohol and tobacco use, occupation, medical history, and family cancer history. These data have been examined extensively in relation to cancer mortality. Vital status of study participants is updated biennially. Cause of death has been documented for over 98% of all deaths that have occurred. Mortality follow-up of the CPS-II participants is complete through 2002 and is expected to continue for many years. (Source: American Cancer Society)

OBJECTIVE 2, PAGE 3

24) It is not always possible to conduct an experiment. Explain why we could not conduct an experiment to investigate the perceived link between high tension wires and leukemia (on humans).

OBJECTIVE 2, PAGE 6

25) There is no point in reinventing the wheel. List some agencies that regularly collect data that are available to the public.

OBJECTIVE 2, PAGE 7

26) What is a census?

27) Why is the U.S. Census so important?

## Section 1.3
## Simple Random Sampling

**Objective**

❶ Obtain a Simple Random Sample

---

INTRODUCTION, PAGE 1

Observational studies can be conducted by administering a survey. When administering a survey, the researcher must first identify the population that is to be targeted.

1) Define: Random sampling

For the results of a survey to be reliable, the characteristics of the individuals in the sample must be representative of the characteristics of the individuals in the population.

The key to obtaining a sample representative of a population is to let chance or randomness play a role in dictating which individuals are in the sample, rather than convenience.

If convenience is used to obtain a sample, the results of the survey are meaningless.

INTRODUCTION, PAGE 2

2) Why are the survey results from the sample taken outside Fenway Park not likely to be reliable?

3) Why are the results of a survey of students in your statistics class likely to be misleading when trying to determine what proportion of students on your campus work?

INTRODUCTION, PAGE 3

4) List the four basic sampling techniques.

## *Objective 1: Obtain a Simple Random Sample*

<u>OBJECTIVE 1, PAGE 1</u>
5) What is a simple random sample?

The number of individuals in the sample is always less than the number of individuals in the population.

<u>OBJECTIVE 1, PAGE 2</u>

---

**Example 1**     *Illustrating Simple Random Sampling*

Sophie has four tickets to a concert. Six of her friends, Yolanda, Michael, Kevin, Marissa, Annie, and Katie, have all expressed an interest in going to the concert. Sophie decides to randomly select three of her six friends to attend the concert.

A) List all possible samples of size $n = 3$ from the population of size $N = 6$. Once an individual is chosen, he/she cannot be chosen again.

B) Comment on the likelihood of the sample containing Michael, Kevin, and Marissa.

---

<u>OBJECTIVE 1, PAGE 5</u>
How do we select the individuals in a simple random sample?
Typically, each individual in the population is assigned a unique number between 1 and $N$, where $N$ is the size of the population. Then $n$ distinct random numbers are selected, where $n$ is the size of the sample. To number the individuals in the population, we need a frame– a list of all the individuals within the population.

OBJECTIVE 1, PAGE 6

 *Answer the following after watching the animation.*

6) What is the frame in this animation?

7) Explain why a second sample of 5 students will most likely be different than the first sample of 5 students?

8) Explain why inferences based on samples vary.

OBJECTIVE 1, PAGE 8

**Example 2**     *Obtaining a Simple Random Sample*

The accounting firm of Senese and Associates has grown. To make sure their clients are still satisfied with the services they are receiving, the company decides to send a survey out to a simple random sample of 5 of its 30 clients.

| TABLE 3 | | |
|---|---|---|
| 01. ABC Electric | 11. Fox Studios | 21. R&Q Realty |
| 02. Brassil Construction | 12. Haynes Hauling | 22. Ritter Engineering |
| 03. Bridal Zone | 13. House of Hair | 23. Simplex Forms |
| 04. Casey's Glass House | 14. John's Bakery | 24. Spruce Landscaping |
| 05. Chicago Locksmith | 15. Logistics Management, Inc. | 25. Thors, Robert DDS |
| 06. DeSoto Painting | 16. Lucky Larry's Bistro | 26. Travel Zone |
| 07. Dino Jump | 17. Moe's Exterminating | 27. Ultimate Electric |
| 08. Euro Car Care | 18. Nick's Tavern | 28. Venetian Gardens Restaurant |
| 09. Farrell's Antiques | 19. Orion Bowling | 29. Walker Insurance |
| 10. First Fifth Bank | 20. Precise Plumbing | 30. Worldwide Wireless |

## Section 1.4
## Other Effective Sampling Methods

**Objectives**

❶ Obtain a Stratified Sample

❷ Obtain a Systematic Sample

❸ Obtain a Cluster Sample

---

INTRODUCTION, PAGE 1

1) What is the goal of sampling?

---

### Objective 1: Obtain a Stratified Sample

OBJECTIVE 1, PAGE 1

2) Explain how to obtain a stratified sample.

OBJECTIVE 1, PAGE 2

| **Example 1** | *Obtaining a Stratified Sample* |
|---|---|

The president of DePaul University wants to conduct a survey to determine the community's opinion regarding campus safety. The president divides the DePaul community into three groups: resident students, nonresident (commuting) students, and staff (including faculty) so that he can obtain a stratified sample.

Suppose there are 6,204 resident students, 13,304 nonresident students, and 2,401 staff, for a total of 21,909 individuals in the population. What percent of the DePaul community is made up of each group?

The president wants to obtain a sample of size 100, with the number of individuals selected from each stratum weighted by the population size. How many individuals should be selected from each stratum?

To obtain the stratified sample, construct a simple random sample within each group.

## *Objective 2: Obtain a Systematic Sample*

OBJECTIVE 2, PAGE 1
3) Explain how to obtain a systematic sample.

**Note:** Because systematic sampling does not require a frame, it is a useful technique when you cannot gather a list of the individuals in the population.

OBJECTIVE 2, PAGE 2

| |
|---|
| **Example 2**      *Obtaining a Systematic Sample without a Frame*<br>The manager of Kroger Food Stores wants to measure the satisfaction of the store's customers. Design a sampling technique that can be used to obtain a sample of 40 customers. |

OBJECTIVE 2, PAGE 4
*Answer the following after watching the video.*
4) What can result from choosing a value of $k$ that is too small?

5) What can result from choosing a value of $k$ that is too large?

OBJECTIVE 2, PAGE 5
*Answer the following after watching the second video after Example 2.*
6) Explain how to determine the value of $k$ if the population size $N$ is known.

7) List the five steps in obtaining a systematic sample.

Step 1

Step 2

Step 3

Step 4

Step 5

---

## *Objective 3: Obtain a Cluster Sample*

8) What is a cluster sample?

OBJECTIVE 3, PAGE 2

---

**Example 3**     *Obtaining a Cluster Sample*

A sociologist wants to gather data regarding household income within the city of Boston. Obtain a sample using cluster sampling.

---

OBJECTIVE 3, PAGE 3

9) If the clusters have homogeneous individuals, is it better to have more clusters with fewer individuals in each cluster or fewer clusters with more individuals in each cluster?

10) If the clusters have heterogeneous individuals, is it better to have more clusters with fewer individuals in each cluster or fewer clusters with more individuals in each cluster?

OBJECTIVE 3, PAGE 5

11) Define: Convenience sampling

OBJECTIVE 3, PAGE 6

**Note:** The most popular convenience samples are those in which the individuals in the sample are self-selected, meaning the individuals themselves decide to participate in the survey.  Self-selected surveys are also called voluntary response samples.

OBJECTIVE 3, PAGE 7

12) Define: Multistage sampling

13) List the two stages Nielsen Media Research uses to investigate TV viewing habits.

OBJECTIVE 3, PAGE 8

14) How many stages does the Census Bureau use for the Current Population Survey? What are those stages?

OBJECTIVE 3, PAGE 9

Researchers need to know how many individuals they must survey to draw conclusions about the population within some predetermined margin of error. They must find a balance between the reliability of the results and the cost of obtaining these results. The bottom line is that time and money determine the level of confidence researchers will place on the conclusions drawn from the sample data. The more time and money researchers have available, the more accurate the results of the statistical inference.

OBJECTIVE 3, PAGE 10

*Watch the animation for a summary of simple random sampling, systematic sampling, stratified sampling, and cluster sampling.*

## Section 1.5
## Bias in Sampling

**Objective**

❶ Explain the Sources of Bias in Sampling

---

### Objective 1: Explain the Sources of Bias in Sampling

<u>OBJECTIVE 1, PAGE 1</u>

1) Define: Bias

2) List the three sources of bias in sampling:

- 

- 

- 

<u>OBJECTIVE 1, PAGE 2</u>

*Answer the following after watching the video.*

3) What is sampling bias?

4) Does a convenience sample have sampling bias?

5) What is undercoverage?

*Answer the following after watching the video.*

6) When does nonresponse bias exist?

7) List two causes of nonresponse bias.

8) List one tool that can be used to control nonresponse bias?

*Answer the following after watching the video.*
9) Under what conditions does response bias exist?

**Note:** Response bias can occur through interviewer error, misrepresented answers, wording of questions, ordering of questions or words, type of question, or data-entry error.

**Note:** An open question allows the respondent to choose his or her response (free response).

**Note:** A closed question requires the respondent to choose from a list of predetermined responses (multiple choice).

OBJECTIVE 1, PAGE 7
**Note: Can a Census Have Bias?**
A question on a census form could be misunderstood, thereby leading to response bias in the results. It is often difficult to contact each individual in a population. For example, the U.S. Census Bureau is challenged to count each homeless person in the country, so the census data published by the U.S. government likely suffers from nonresponse bias.

OBJECTIVE 1, PAGE 8
Define the following terms.

10) Nonsampling Error:

11) Sampling error:

## Section 1.6
## The Design of Experiments

**Objectives**

❶ Describe the Characteristics of an Experiment

❷ Explain the Steps in Designing an Experiment

❸ Explain the Completely Randomized Design

❹ Explain the Matched-Pairs Design

INTRODUCTION, PAGE 1

*Watch the video for a review of the language used in observational studies.*
Review the definitions of cross-sectional studies, case-control studies, and cohort studies.

- In observational studies, we cannot make statements of *causality* between the explanatory variable(s) and the response variable.

- The response variable measures the outcome of the study.

- The explanatory variable is the variable whose impact we want to see has on the response variable.

### *Objective 1: Describe the Characteristics of an Experiment*

OBJECTIVE 1, PAGE 1

*Define the following terms after watching the video.*
1) Experiment:

2) Factor:

3) Treatment:

*Define the following terms after watching the video.*

4) Experimental unit:

5) Control group:

6) Placebo:

7) Blinding:

8) Single-blind:

9) Double-blind:

OBJECTIVE 1, PAGE 2

The use of placebos in designed experiments is a way to form a control group in a designed experiment.

10) What is the placebo effect?

OBJECTIVE 1, PAGE 3

Recall confounding in a study occurs when the effects of two or more explanatory variables are not separated. In designed experiments, confounding may occur as a result of a confounding variable, which is an explanatory variable that was considered in a study whose effect cannot be distinguished from a second explanatory variable in the study.

OBJECTIVE 1, PAGE 6

**Example 1**    *The Characteristics of an Experiment*

Lipitor is a cholesterol-lowering drug made by Pfizer. In the Collaborative Atorvastatin Diabetes Study (CARDS), the effect of Lipitor on cardiovascular disease was assessed in 2838 subjects, ages 40 to 75, with type 2 diabetes, without prior history of cardiovascular disease. In this placebo-controlled, double-blind experiment, subjects were randomly allocated to either Lipitor 10 mg daily (1428) or placebo (1410) and were followed for 4 years. The response variable whether there was an occurrence of any major cardiovascular event or not.

Lipitor significantly reduced the rate of major cardiovascular events (83 events in the Lipitor group versus 127 events in the placebo group). There were 61 deaths in the Lipitor group versus 82 deaths in the placebo group.

A) What does it mean for the experiment to be placebo-controlled?

B) What does it mean for the experiment to be double-blind?

C) What is the population for which this study applies? What is the sample?

D) What are the treatments?

E) What is the response variable? Is it qualitative or quantitative?

---

## *Objective 2: Explain the Steps in Designing an Experiment*

<u>OBJECTIVE 2, PAGE 1</u>
**Steps in Conducting a Designed Experiment**
Fill in each step.

*Step 1*: _____

The statement of the problem should be as explicit as possible and should provide the experimenter with direction. The statement must also identify the response variable and the population to be studied. Often, the statement is referred to as the *claim*.

*Step 2*: _____

The factors are usually identified by an expert in the field of study. In identifying the factors, ask, "What things affect the value of the response variable?" After the factors are identified, determine which factors to fix at some predetermined level, which to manipulate, and which to leave uncontrolled.

*Step 3*: _____

As a general rule, choose as many experimental units as time and money allow. Techniques exist for determining sample size, provided certain information is available.

OBJECTIVE 2, PAGE 1 (CONTINUED)

*Step 4:* _____

Factors can be dealt with in two ways - control or randomize.
Control means to either set the factor at one value throughout the experiment or set the level of the factor at various levels).
Randomize means to randomly assign the experimental units to various treatment groups.

*Step 5:* _____

Replication occurs when each treatment is applied to more than one experimental unit.

*Step 6:* _____

Inferential statistics is a process in which generalizations about a population are made on the basis of results obtained from a sample.

OBJECTIVE 2, PAGE 2
List the six steps for the Lipitor study in Example 1 (Objective 1, Page 6)
**Step 1:** *Identify the Problem to be Solved*

**Step 2:** *Determine the Factors That Affect the Response Variable*

**Step 3:** *Determine the Number of Experimental Units*

**Step 4:** *Determine the Level of Each Factor*

**Step 5:** *Conduct the Experiment*

**Step 6:** *Test the Claim*

## *Objective 3: Explain the Completely Randomized Design*

OBJECTIVE 3, PAGE 1
11) What is a completely randomized design?

OBJECTIVE 3, PAGE 2

**Example 2**     *A Completely Randomized Design*

A farmer wishes to determine the optimal level of a new fertilizer on his soybean crop. Design an experiment that will assist him.

OBJECTIVE 3, PAGE 3
Sketch the experimental design from Example 2 (Objective 3, Page 2).

12) Explain why this experimental design is a completely randomized design.

## Objective 4: Explain the Matched-Pairs Design

OBJECTIVE 4, PAGE 1

13) What is a matched-pairs design?

The pairs are selected so that they are related in some way.

There are only two levels of treatment in a matched-pairs design.

OBJECTIVE 4, PAGE 2

**Example 3**    *A Matched-Pairs Design*

An educational psychologist wants to determine whether listening to music has an effect on a student's ability to learn. Design an experiment to help the psychologist answer the question.

# Chapter 2 – Organizing and Summarizing Data

## OUTLINE

**Putting It Together**

Chapter 1 discussed how to identify the research objective and collect data. We learned that data can be obtained from either observational studies or designed experiments. When data are obtained, they are referred to as **raw data**.

The purpose of this chapter is to learn how to organize raw data into a meaningful form so that we can understand what the data are telling us. The first step in determining how to organize raw data is to determine whether the data is qualitative or quantitative.

## Section 2.1
## Organizing Qualitative Data

**Objectives**

❶ Organize Qualitative Data in Tables

❷ Construct Bar Graphs

❸ Construct Pie Charts

---

### Objective 1: Organize Qualitative Data in Tables

OBJECTIVE 1, PAGE 1

1) What is used to list each category of data and the number of occurrences for each category of data?

OBJECTIVE 1, PAGE 2

| **Example 1** | **Organizing Qualitative Data into a Frequency Distribution** |
| --- | --- |

A physical therapist wants to determine types of rehabilitation required by her patients. To do so, she obtains a simple random sample of 30 of her patients and records the body part requiring rehabilitation. (See Table 1.) Construct a frequency distribution of location of injury.

**Table 1**

| Back | Back | Hand | Wrist | Back | Back |
| Groin | Elbow | Back | Back | Back | Groin |
| Shoulder | Shoulder | Hip | Knee | Hip | Shoulder |
| Neck | Knee | Knee | Shoulder | Shoulder | Neck |
| Back | Back | Back | Back | Knee | Back |

Data from Krystal Catton, student at Joliet Junior College

OBJECTIVE 1, PAGE 3

2) In any frequency distribution, it is a good idea to add up the frequency column. What should the total be equal to?

OBJECTIVE 1, PAGE 6

3) Define the relative frequency of a category.

4) What is a relative frequency distribution?

OBJECTIVE 1, PAGE 7

| **Example 2** | ***Constructing a Relative Frequency Distribution of Qualitative Data*** |
|---|---|

Using the summarized data in Table 2, construct a relative frequency distribution.

**Table 2**

| Body Part | Frequency |
|-----------|-----------|
| Back | 12 |
| Hand | 2 |
| Wrist | 2 |
| Groin | 1 |
| Elbow | 1 |
| Shoulder | 4 |
| Hip | 2 |
| Knee | 5 |
| Neck | 1 |

<u>OBJECTIVE 1, PAGE 8</u>
5) When working with a relative frequency distribution, what should the total of the relative frequencies be equal to? Why?

---

## Objective 2: Construct Bar Graphs

<u>OBJECTIVE 2, PAGE 1</u>
6) Explain how a bar graph is constructed. What do the heights of each rectangle represent?

<u>OBJECTIVE 2, PAGE 2</u>

| Example 3 | *Constructing a Frequency and Relative Frequency Bar Graph* |
|---|---|

Use the data summarized in Table 3 to construct a frequency bar graph and relative frequency bar graph.

**Table 3**

| Body Part | Frequency | Relative Frequency |
|---|---|---|
| Back | 12 | 0.4 |
| Hand | 2 | 0.0667 |
| Wrist | 2 | 0.0667 |
| Groin | 1 | 0.0333 |
| Elbow | 1 | 0.0333 |
| Shoulder | 4 | 0.1333 |
| Hip | 2 | 0.0667 |
| Knee | 5 | 0.1667 |
| Neck | 1 | 0.0333 |

OBJECTIVE 2, PAGE 4

7) What is a Pareto chart?

OBJECTIVE 2, PAGE 5

8) Explain why it is best to use relative frequencies when comparing data sets.

OBJECTIVE 2, PAGE 6

| **Example 4** | ***Comparing Two Data Sets*** |
|---|---|

The frequency data in Table 4 represent the educational attainment (level of education) in 1990 and 2016 of adults 25 years and older who are U.S. residents. The data are in thousands. So 39,344 represents 39,344,000.

**Table 4**

| Educational Attainment | 1990 | 2016 |
|---|---|---|
| Not a high school graduate | 39,344 | 23,453 |
| High school diploma | 47,643 | 62,002 |
| Some college, no degree | 29,780 | 36,003 |
| Associate's degree | 9792 | 21,657 |
| Bachelor's degree | 20,833 | 44,778 |
| Graduate or professional degree | 11,478 | 27,122 |
| Totals | 158,870 | 215,015 |

A) Draw a side-by-side relative frequency bar graph of the data.

OBJECTIVE 2, PAGE 6 (CONTINUED)

B) The side-by-side relative frequency bar graph shows additional information that was not easy to identify from the frequency table in Table 4. Comment on the interesting features of the side-by-side relative frequency bar graph.

OBJECTIVE 2, PAGE 8

9) Explain when it would be preferable to use horizontal bars rather than vertical bars when constructing a bar graph.

## *Objective 3: Construct Pie Charts*

OBJECTIVE 3, PAGE 1

10) What is a pie chart?

<u>Objective 3, Page 2</u>

---

**Example 5    *Constructing a Pie Chart***

The frequency data presented in Table 6 represent the educational attainment of U.S. residents 25 years and older in 2016. The data are in thousands so 23,453 represents 23,453,000. Construct a pie chart of the data.

**Table 6**

| Educational Attainment | 2016 |
|---|---|
| Not a high school graduate | 23,453 |
| High school diploma | 62,002 |
| Some college, no degree | 36,003 |
| Associate's degree | 21,657 |
| Bachelor's degree | 44,778 |
| Graduate or professional degree | 27,122 |
| Totals | 215,015 |

---

<u>Objective 3, Page 5</u>

*Answer the following after watching the video.*

11) Which graph, a pie chart or a bar graph, is better at comparing one category to another category?

12) Which graph, a pie chart or a bar graph, is better at comparing one category to the whole?

## Section 2.2
## Organizing Quantitative Data: The Popular Displays

**Objectives**

❶ Organize Discrete Data in Tables

❷ Construct Histograms of Discrete Data

❸ Organize Continuous Data in Tables

❹ Construct Histograms of Continuous Data

❺ Draw Dot Plots

❻ Identify the Shape of a Distribution

---

### *Objective 1: Organize Discrete Data in Tables*

OBJECTIVE 1, PAGE 1

1) What do we use to create the classes when the number of distinct data values of a discrete variable is small?

OBJECTIVE 1, PAGE 2

| Example 1 | *Constructing Frequency and Relative Frequency Distributions from Discrete Data* |
| --- | --- |

The manager of a Wendy's® fast-food restaurant wants to know the typical number of customers who arrive during the lunch hour. The data represent the number of customers who arrive at Wendy's for 40 randomly selected 15-minute intervals of time during lunch.
Construct a frequency and relative frequency distribution.

**Number of Arrivals at Wendy's**

| | | | | | | | |
| --- | --- | --- | --- | --- | --- | --- | --- |
| 7 | 6 | 6 | 6 | 4 | 6 | 2 | 6 |
| 5 | 6 | 6 | 11 | 4 | 5 | 7 | 6 |
| 2 | 7 | 1 | 2 | 4 | 8 | 2 | 6 |
| 6 | 5 | 5 | 3 | 7 | 5 | 4 | 6 |
| 2 | 2 | 9 | 7 | 5 | 9 | 8 | 5 |

*Objective 2: Construct Histograms of Discrete Data*

<u>OBJECTIVE 2, PAGE 1</u>

2) Explain how a histogram is constructed.

<u>OBJECTIVE 2, PAGE 2</u>

**Example 2**    *Drawing a Histogram of Discrete Data*

Construct a frequency histogram and a relative frequency histogram using the data in Table 9. Recall that this table summarizes the data for the number of customers who arrive at Wendy's for 40 randomly selected 15-minute intervals of time during lunch.

**Table 9**

| Number of Customers | Frequency | Relative Frequency |
|---|---|---|
| 1 | 1 | 0.025 |
| 2 | 6 | 0.15 |
| 3 | 1 | 0.025 |
| 4 | 4 | 0.1 |
| 5 | 7 | 0.175 |
| 6 | 11 | 0.275 |
| 7 | 5 | 0.125 |
| 8 | 2 | 0.05 |
| 9 | 2 | 0.05 |
| 10 | 0 | 0.0 |
| 11 | 1 | 0.025 |

## Objective 3: Organize Continuous Data in Tables

OBJECTIVE 3, PAGE 1

**Note:** When a data set consists of a large number of different discrete data values or when a data set consists of continuous data, we must create classes by using intervals of numbers.

Define the following terms.

3) Lower class limit

4) Upper class limit

5) Class width

6) When creating classes for a frequency distribution, the classes must not _____.

OBJECTIVE 3, PAGE 2

7) What is an open-ended table?

**Example 3**   *Organizing Continuous Data into a Frequency and Relative Frequency Distribution*

Suppose you are considering investing in a Roth IRA. You collect the data in Table 12, which represent the five-year rate of return (in percent, adjusted for sales charges) for a simple random sample of 40 large-blend mutual funds. Construct a frequency and relative frequency distribution of the data.

**Table 12**
**Five-Year Rate of Return of Mutual Funds (in percent)**

| | | | | | | | |
|---|---|---|---|---|---|---|---|
| 10.94 | 14.60 | 12.80 | 16.00 | 11.93 | 15.68 | 9.03 | 13.40 |
| 10.53 | 13.98 | 13.86 | 12.36 | 13.54 | 9.94 | 13.93 | 13.63 |
| 14.12 | 14.88 | 14.77 | 13.13 | 8.28 | 19.43 | 12.98 | 13.16 |
| 12.26 | 14.20 | 14.80 | 13.26 | 13.67 | 10.08 | 14.86 | 8.71 |
| 12.17 | 10.26 | 15.22 | 13.36 | 13.55 | 13.90 | 15.64 | 12.80 |

Data from Morningstar.com

---

## Objective 4: Construct Histograms of Continuous Data

**Example 4**   *Drawing a Histogram of Continuous Data*

Construct a frequency and relative frequency histogram of the five-year rate of return data discussed in Example 3.

**Table 12**
**Five-Year Rate of Return of Mutual Funds (in percent)**

| | | | | | | | |
|---|---|---|---|---|---|---|---|
| 10.94 | 14.60 | 12.80 | 16.00 | 11.93 | 15.68 | 9.03 | 13.40 |
| 10.53 | 13.98 | 13.86 | 12.36 | 13.54 | 9.94 | 13.93 | 13.63 |
| 14.12 | 14.88 | 14.77 | 13.13 | 8.28 | 19.43 | 12.98 | 13.16 |
| 12.26 | 14.20 | 14.80 | 13.26 | 13.67 | 10.08 | 14.86 | 8.71 |
| 12.17 | 10.26 | 15.22 | 13.36 | 13.55 | 13.90 | 15.64 | 12.80 |

Data from Morningstar.com

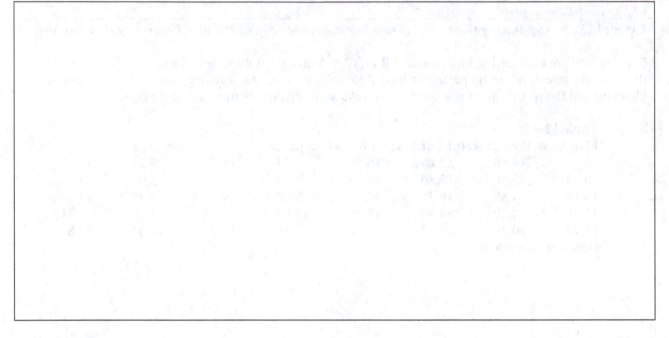

<u>OBJECTIVE 4, PAGE 4</u>

There is no one correct frequency distribution for a particular set of data. However, some frequency distributions better illustrate patterns within the data than others. So constructing frequency distributions is somewhat of an art form. Use the distribution that seems to provide the best overall summary of the data.

<u>OBJECTIVE 4, PAGE 5</u>

*Answer the following after using the applet in Activity 1: Choosing Class Width.*
8) What happens to the number of classes as the bin width increases?

<u>OBJECTIVE 4, PAGE 7</u>

9) The number of classes in a frequency distribution is typically between what two numbers?

10) Explain how to choose the lower class limit of the first class in a frequency distribution.

11) Once you decide on the number of classes, explain how to determine the class width.

## *Objective 5: Draw Dot Plots*

OBJECTIVE 5, PAGE 1

12) Explain how to draw a dot plot.

OBJECTIVE 5, PAGE 2

**Example 5**   *Drawing a Dot Plot*

Draw a dot plot for the data from Table 8.

**Table 8**
**Number of Arrivals at Wendy's**

| | | | | | | | |
|---|---|---|---|---|---|---|---|
| 7 | 6 | 6 | 6 | 4 | 6 | 2 | 6 |
| 5 | 6 | 6 | 11 | 4 | 5 | 7 | 6 |
| 2 | 7 | 1 | 2 | 4 | 8 | 2 | 6 |
| 6 | 5 | 5 | 3 | 7 | 5 | 4 | 6 |
| 2 | 2 | 9 | 7 | 5 | 9 | 8 | 5 |

***Objective 6: Identify the Shape of a Distribution***

<u>OBJECTIVE 6, PAGE 1</u>

13) Draw an example of a uniform distribution.

14) Draw an example of a bell-shaped distribution.

15) Draw an example of a distribution that is skewed right.

<u>OBJECTIVE 6, PAGE 1 (CONTINUED)</u>

16) Draw an example of a distribution that is skewed left.

<u>OBJECTIVE 6, PAGE 2</u>

**Example 6**    *Identifying the Shape of a Distribution*

Figure 10 displays the histogram obtained in Example 4 for the five-year rate of return for large-blended mutual funds. Describe the shape of the distribution.

**Figure 10**
**Five-Year Rate of Return for Large
Blend Mutual Funds**

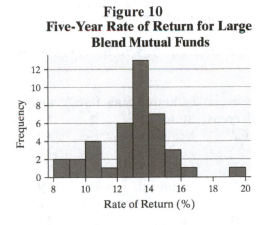

# Section 2.3
## Additional Displays of Quantitative Data

**Objectives**

❶ Draw Stem-and-Leaf Plots

❷ Construct Frequency Polygons

❸ Create Cumulative Frequency and Relative Frequency Distributions

❹ Construct Frequency and Relative Frequency Ogives

❺ Draw Time-Series Graphs

---

## *Objective 1: Draw Stem-and-Leaf Plots*

OBJECTIVE 1, PAGE 1

1) In a stem-and-leaf plot, how are the stem and leaf identified?

OBJECTIVE 1, PAGE 2

| **Example 1** | **Constructing a Stem-and-Leaf Plot** |

The data in Table 14 represent the percentage of persons living in poverty, by state, in 2015. Draw a stem-and-leaf plot of the data.

**Table 14**

**Percentage of People in Poverty by State Using Two-Year Averages: 2011–2012**

| State | Percent | State | Percent | State | Percent |
|---|---|---|---|---|---|
| Alabama | 16.3 | Kentucky | 19.5 | North Dakota | 10.7 |
| Alaska | 9.2 | Louisiana | 18.6 | Ohio | 13.6 |
| Arizona | 17.2 | Maine | 12.3 | Oklahoma | 14.2 |
| Arkansas | 16.1 | Maryland | 9.6 | Oregon | 11.9 |
| California | 13.9 | Massachusetts | 11.5 | Pennsylvania | 12.3 |
| Colorado | 9.9 | Michigan | 12.8 | Rhode Island | 11.8 |
| Connecticut | 9.1 | Minnesota | 7.8 | South Carolina | 14.3 |
| Delaware | 11.1 | Mississippi | 19.1 | South Dakota | 13.9 |
| District of Columbia | 16.6 | Missouri | 9.8 | Tennessee | 14.7 |
| Florida | 16.2 | Montana | 11.9 | Texas | 14.7 |
| Georgia | 18.1 | Nebraska | 10.3 | Utah | 9.3 |
| Hawaii | 10.9 | Nevada | 13.0 | Vermont | 10.7 |
| Idaho | 12.3 | New Hampshire | 7.3 | Virginia | 10.9 |
| Illinois | 10.9 | New Jersey | 11.2 | Washington | 11.4 |
| Indiana | 13.5 | New Mexico | 19.7 | West Virginia | 14.5 |
| Iowa | 10.4 | New York | 14.2 | Wisconsin | 11.4 |
| Kansas | 14.2 | North Carolina | 15.3 | Wyoming | 9.8 |

Data from united States Census Bureau

OBJECTIVE 1, PAGE 2 (CONTINUED)

OBJECTIVE 1, PAGE 3

2) List the four steps for constructing a stem-and-leaf plot.

OBJECTIVE 1, PAGE 4

3) List an advantage that a stem-and-leaf plot has over frequency distributions and histograms.

4) Under what conditions do stem-and-leaf plots lose their usefulness?

5) When constructing a stem-and-leaf plot, under what conditions is it advisable to use split stems?

---

## Objective 2: Construct Frequency Polygons

6) Explain how to construct a frequency polygon.

**Example 2**    *Constructing a Frequency Polygon*

Draw a frequency polygon of the five-year rate of return data listed in Table 16.

**Table 16**
**Five-Year Rate of Return of Mutual Funds (in percent)**

| | | | | | | | |
|---|---|---|---|---|---|---|---|
| 10.94 | 14.60 | 12.80 | 16.00 | 11.93 | 15.68 | 9.03 | 13.40 |
| 10.53 | 13.98 | 13.86 | 12.36 | 13.54 | 9.94 | 13.93 | 13.63 |
| 14.12 | 14.88 | 14.77 | 13.13 | 8.28 | 19.43 | 12.98 | 13.16 |
| 12.26 | 14.20 | 14.80 | 13.26 | 13.67 | 10.08 | 14.86 | 8.71 |
| 12.17 | 10.26 | 15.22 | 13.36 | 13.55 | 13.90 | 15.64 | 12.80 |

Data from Morningstar.com

## Objective 3: Create Cumulative Frequency and Relative Frequency Distributions

OBJECTIVE 3, PAGE 1

7) What does a cumulative frequency distribution display?

8) What does a cumulative relative frequency distribution display?

9) Explain how to find the cumulative frequency for the fifth class in a cumulative frequency distribution.

OBJECTIVE 3, PAGE 2

| Example 3 | Constructing a Cumulative and Cumulative Relative Frequency Distribution |

Obtain a cumulative frequency distribution and cumulative relative frequency distribution for the five-year rate of return data listed in Table 13.

| Class (5-year rate of return) | Frequency | Relative Frequency |
|---|---|---|
| 8–8.99 | 2 | 0.05 |
| 9–9.99 | 2 | 0.05 |
| 10–10.99 | 4 | 0.10 |
| 11–11.99 | 1 | 0.025 |
| 12–12.99 | 6 | 0.15 |
| 13–13.99 | 13 | 0.325 |
| 14–14.99 | 7 | 0.175 |
| 15–15.99 | 3 | 0.075 |
| 16–16.99 | 1 | 0.025 |
| 17–17.99 | 0 | 0 |
| 18–18.99 | 0 | 0 |
| 19–19.99 | 1 | 0.025 |

## Objective 4: Construct Frequency and Relative Frequency Ogives

OBJECTIVE 4, PAGE 1

10) What does an ogive represent?

11) Explain the difference between *x*-coordinates for a frequency polygon and a frequency ogive.

12) Explain the difference between *y*-coordinates for a frequency polygon and a frequency ogive.

**Example 4**     *Constructing Ogives*

Draw a relative frequency ogive of the five-year rate of return data listed in Table 17.

**Table 17**

| Class (5-year rate of return) | Frequency | Relative Frequency | Cumulative Frequency | Cumulative Relative Frequency |
|---|---|---|---|---|
| 8–8.99 | 2 | 0.05 | 2 | 0.05 |
| 9–9.99 | 2 | 0.05 | 4 | 0.1 |
| 10–10.99 | 4 | 0.10 | 8 | 0.2 |
| 11–11.99 | 1 | 0.025 | 9 | 0.225 |
| 12–12.99 | 6 | 0.15 | 15 | 0.375 |
| 13–13.99 | 13 | 0.325 | 28 | 0.7 |
| 14–14.99 | 7 | 0.175 | 35 | 0.875 |
| 15–15.99 | 3 | 0.075 | 38 | 0.95 |
| 16–16.99 | 1 | 0.025 | 39 | 0.975 |
| 17–17.99 | 0 | 0 | 39 | 0.975 |
| 18–18.99 | 0 | 0 | 39 | 0.975 |
| 19–19.99 | 1 | 0.025 | 40 | 1 |

*Objective 5: Draw Time-Series Graphs*

13) Define time-series data.

14) Explain how to create a time-series plot.

**Example 5**     *Drawing a Time-Series Plot*

The Partisan Conflict Index (PCI) tracks the degrees of political disagreement among U.S. politicians in the federal government. It is found by measuring the frequency of newspaper articles reporting disagreement in a given month. Higher values of the index suggest greater conflict among political parties, Congress, and the President. The data in Table 18 represent the PCI in March from 1999 to 2017. Construct a time-series plot of the data. In what year was the index highest? In what year was the index lowest?

**Table 18**

| Year | Partisan Conflict Index (PCI) | Year | Partisan Conflict Index (PCI) |
|------|-------------------------------|------|-------------------------------|
| 1999 | 85.87 | 2009 | 88.04 |
| 2000 | 94.67 | 2010 | 142.42 |
| 2001 | 78.23 | 2011 | 155.83 |
| 2002 | 86.67 | 2012 | 154.18 |
| 2003 | 88.49 | 2013 | 180.56 |
| 2004 | 98.55 | 2014 | 131.4 |
| 2005 | 100.07 | 2015 | 163.54 |
| 2006 | 91.49 | 2016 | 173.88 |
| 2007 | 85.44 | 2017 | 270.72 |
| 2008 | 90.87 | | |

*Source*: Federal Reserve Bank of Philadelphia

## Section 2.4
## Graphical Misrepresentations of Data

**Objective**

❶ Describe What Can Make a Graph Misleading or Deceptive

---

### Objective 1: Describe What Can Make a Graph Misleading or Deceptive

OBJECTIVE 1, PAGE 1

*Answer the following after watching the video.*

1) Explain the difference between a graph that is misleading and a graph that is deceiving.

2) List what the most common misrepresentations of data involve.

- Increments between tick marks should be consistent.
- Scales for comparative graphs should be the same.
- The baseline, or zero point, should be at the bottom of the graph.

OBJECTIVE 1, PAGE 2

| Example 1 | *Misrepresentations of Data* |

A home security company located in Minneapolis, Minnesota, develops a summer ad campaign with the slogan "When you leave for vacation, burglars leave for work." According to the city of Minneapolis, roughly 20% of home burglaries occur during the peak vacation months of July and August. The advertisement contains the graphic shown. Explain what is wrong with the graphic.

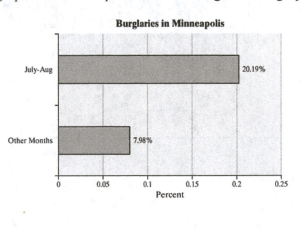

Chapter 2: Organizing and Summarizing Data

| Example 2 | *Misrepresentations of Data by Manipulating the Vertical Scale* |

A national news organization developed the following graphic to illustrate the change in the highest marginal tax rate effective January 1, 2013. Why might this graph be considered misleading?

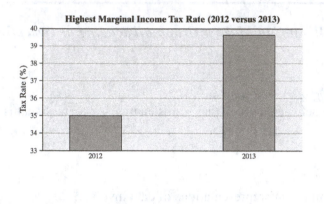

| Example 3 | *Misrepresentations of Data by Manipulating the Vertical Scale* |

The graph depicts the number of residents in the United States living in poverty. Why might this graph be considered misrepresentative?

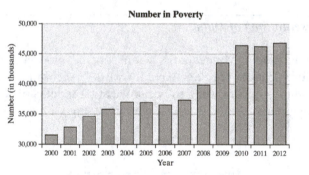

Copyright © 2019 Pearson Education, Inc.

<u>OBJECTIVE 1, PAGE 7</u>

**Example 4**    *Misrepresentations of Data*

The bar graph shown is a *USA Today*-type graph. A survey was conducted by Impulse Research in which individuals were asked how they would flush a toilet when the facilities are not sanitary. What is wrong with the graphic?

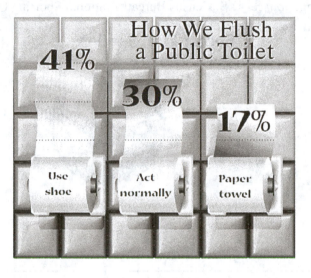

<u>OBJECTIVE 1, PAGE 9</u>

3) Why is the use of 3-D effects strongly discouraged?

4) Why do we emphasize that the bars or classes should have the same width?

| Example 5 | *Misrepresentations of Data by Manipulating Dimension* |

Soccer continues to grow in popularity as a sport in the United States. In 1991, there were approximately 10 million participants in the United States aged 7 years and older. By 2009, this number had climbed to 14 million. To illustrate this increase, we could create a graphic like the one shown below. Describe how the graph may be misleading. *Source:* U.S. Census Bureau; National Sporting Goods Association

**Soccer Participation**

1991                    2009

| Example 6 | *Misrepresentations of Data: Three-Dimensional Scale* |

The figure represents the educational attainment (level of education) in 2016 of adults 25 years and older who are U.S. residents. Why might this graph be considered misrepresentative?

**Educational Attainment, 2012**

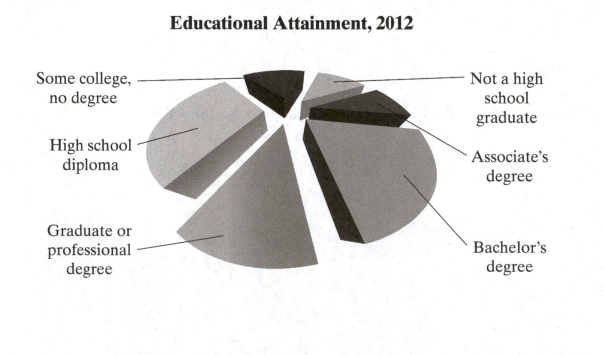

Some college, no degree

High school diploma

Graduate or professional degree

Not a high school graduate

Associate's degree

Bachelor's degree

# Chapter 3 – Numerically Summarizing Data

## Putting It Together

When we look at a distribution of data, we should consider three characteristics of the distribution: shape, center, and spread. In the last chapter, we discussed methods for organizing raw data into tables and graphs. These graphs (such as the histogram) allow us to identify the shape of the distribution: symmetric (in particular, bell shaped or uniform), skewed right, or skewed left.

The center and spread are numerical summaries of the data. The center of a data set is commonly called the average. There are many ways to describe the average value of a distribution. In addition, there are many ways to measure the spread of a distribution. The most appropriate measure of center and spread depends on the distribution's shape.

Once these three characteristics of the distribution are known, we can analyze the data for interesting features, including unusual data values, called outliers.

## Section 3.1
## Measures of Central Tendency

**Objectives**

❶ Determine the Arithmetic Mean of a Variable from Raw Data

❷ Determine the Median of a Variable from Raw Data

❸ Explain What It Means for a Statistic to be Resistant

❹ Determine the Mode of a Variable from Raw Data

*Objective 1: Determine the Arithmetic Mean of a Variable from Raw Data*

INTRODUCTION, PAGE 1

*Answer the following after watching the video.*

1) What does a measure of central tendency describe?

OBJECTIVE 1, PAGE 1

2) Explain how to compute the arithmetic mean of a variable.

3) What symbols are used to represent the population mean and the sample mean?

OBJECTIVE 1, PAGE 2

4) List the formulas used to compute the population mean and the sample mean.

**Note:** Throughout this course, we agree to round the mean to one more decimal place than that in the raw data.

OBJECTIVE 1, PAGE 3

---

**Example 1**    *Computing a Population Mean and a Sample Mean*

Table 1 shows the first exam scores of the ten students enrolled in Introductory Statistics.

| **Table 1** | **Student** | **Score** |
|---|---|---|
| | 1. Michelle | 82 |
| | 2. Ryanne | 77 |
| | 3. Bilal | 90 |
| | 4. Pam | 71 |
| | 5. Jennifer | 62 |
| | 6. Dave | 68 |
| | 7. Joel | 74 |
| | 8. Sam | 84 |
| | 9. Justine | 94 |
| | 10. Juan | 88 |

A) Compute the population mean, $\mu$.

B) Find a simple random sample of size $n = 4$ students.

C) Compute the sample mean, $\bar{x}$, of the sample found in part (B).

---

OBJECTIVE 1, PAGE 5

*Answer the following after experimenting with the fulcrum animation.*

5) What is the mean of the data?

6) Explain why it is helpful to think of the mean as the center of gravity.

## Objective 2: Determine the Median of a Variable from Raw Data

<u>OBJECTIVE 2, PAGE 1</u>

7) Define the median of a variable.

<u>OBJECTIVE 2, PAGE 2</u>

8) List the three steps in finding the median of a data set.

<u>OBJECTIVE 2, PAGE 3</u>

| **Example 2** | *Determining the Median of a Data Set (Odd Number of Observations)* |
| --- | --- |

Table 2 shows the length (in seconds) of a random sample of songs released in the 1970s. Find the median length of the songs.

**Table 2**

| Song Name | Length |
| --- | --- |
| "Sister Golden Hair" | 201 |
| "Black Water" | 257 |
| "Free Bird" | 284 |
| "The Hustle" | 208 |
| "Southern Nights" | 179 |
| "Stayin' Alive" | 222 |
| "We Are Family" | 217 |
| "Heart of Glass" | 206 |
| "My Sharona" | 240 |

OBJECTIVE 2, PAGE 5

---

**Example 3**     *Determining the Median of a Data Set (Even Number of Observations)*

Find the median score of the data in Table 1.

| Table 1 | Student | Score |
|---|---|---|
| | 1. Michelle | 82 |
| | 2. Ryanne | 77 |
| | 3. Bilal | 90 |
| | 4. Pam | 71 |
| | 5. Jennifer | 62 |
| | 6. Dave | 68 |
| | 7. Joel | 74 |
| | 8. Sam | 84 |
| | 9. Justine | 94 |
| | 10. Juan | 88 |

---

## Objective 3: Explain What It Means for a Statistic to be Resistant

OBJECTIVE 3, PAGE 1

*Answer the following as you work through the Mean versus Median Applet.*

9) When the mean and median are approximately 2, how does adding a single observation near 9 affect the mean? How does it affect the median?

10) When the mean and median are approximately 2, how does adding a single observation near 24 affect the mean? The median?

OBJECTIVE 3, PAGE 1 (CONTINUED)
11) When the mean and median are approximately 40, how does dragging the new observation from 35 toward 0 affect the mean? How does it affect the median?

OBJECTIVE 3, PAGE 2

*Answer the following as you watch the video.*
12) Which measure, the mean or the median, is least affected by extreme observations?

13) Define what it means for a numerical summary of data to be resistant.

14) Which measure, the mean or the median, is resistant?

OBJECTIVE 3, PAGE 3
15) State the reason that we compute the mean.

OBJECTIVE 3, PAGE 7

*Answer the following as you work through Activity 2: Relation among the Mean, Median, and Distribution Shape.*

16) If a distribution is skewed left, what is the relation between the mean and median?

17) If a distribution is skewed right, what is the relation between the mean and median?

<u>OBJECTIVE 3, PAGE 7 (CONTINUED)</u>
18) If a distribution is symmetric, what is the relation between the mean and median?

<u>OBJECTIVE 3, PAGE 11</u>
19) Sketch three graphs showing the relation between the mean and median for distributions that are skewed left, symmetric, and skewed right.

<u>OBJECTIVE 3, PAGE 12</u>

| **Example 4** | ***Describing the Shape of a Distribution*** |
|---|---|

The data in Table 4 represent the birth weights (in pounds) of 50 randomly sampled babies.
A) Find the mean and median birth weight.
B) Describe the shape of the distribution.
C) Which measure of central tendency best describes the average birth weight?

**Table 4**

| 5.8 | 7.4 | 9.2 | 7.0 | 8.5 | 7.6 |
|---|---|---|---|---|---|
| 7.9 | 7.8 | 7.9 | 7.7 | 9.0 | 7.1 |
| 8.7 | 7.2 | 6.1 | 7.2 | 7.1 | 7.2 |
| 7.9 | 5.9 | 7.0 | 7.8 | 7.2 | 7.5 |
| 7.3 | 6.4 | 7.4 | 8.2 | 9.1 | 7.3 |
| 9.4 | 6.8 | 7.0 | 8.1 | 8.0 | 7.5 |
| 7.3 | 6.9 | 6.9 | 6.4 | 7.8 | 8.7 |
| 7.1 | 7.0 | 7.0 | 7.4 | 8.2 | 7.2 |
| 7.6 | 6.7 | | | | |

## Objective 4: Determine the Mode of a Variable from Raw Data

<u>OBJECTIVE 4, PAGE 1</u>
20) Define the mode of a variable.

21) Under what conditions will a set of data have no mode?

22) Under what conditions will a set of data have two modes?

<u>OBJECTIVE 4, PAGE 2</u>

| Example 5 | *Finding the Mode of Quantitative Data* |
|---|---|

The following data represent the number of O-ring failures on the shuttle Columbia for the 17 flights prior to its fatal flight:

0, 0, 0, 0, 0, 0, 0, 0, 0, 0, 0, 1, 1, 1, 1, 2, 3

Find the mode number of O-ring failures.

<u>OBJECTIVE 4, PAGE 3</u>

| Example 6 | *Finding the Mode of Quantitative Data* |
|---|---|

Find the mode of the exam score data listed in Table 1.

| Table 1 | Student | Score |
|---|---|---|
| | 1. Michelle | 82 |
| | 2. Ryanne | 77 |
| | 3. Bilal | 90 |
| | 4. Pam | 71 |
| | 5. Jennifer | 62 |
| | 6. Dave | 68 |
| | 7. Joel | 74 |
| | 8. Sam | 84 |
| | 9. Justine | 94 |
| | 10. Juan | 88 |

<u>OBJECTIVE 4, PAGE 5</u>

23) What does it mean when we say that a data set is bimodal? Multimodal?

<u>OBJECTIVE 4, PAGE 6</u>

| **Example 7** | ***Finding the Mode of Qualitative Data*** |
|---|---|

The data in Table 5 represent the location of injuries that required rehabilitation by a physical therapist. Determine the mode location of injury.

**Table 5**

| Back | Back | Hand | Neck | Knee | Knee |
|---|---|---|---|---|---|
| Wrist | Back | Groin | Shoulder | Shoulder | Back |
| Elbow | Back | Back | Back | Back | Back |
| Back | Shoulder | Shoulder | Knee | Knee | Back |
| Hip | Knee | Hip | Hand | Back | Wrist |

Data from Krystal Catton, student at Joliet Junior College

<u>OBJECTIVE 4, PAGE 8</u>

## *Summary*

24) List the conditions for determining when to use the following measures of central tendency.

A) Mean

B) Median

C) Mode

## Section 3.2
## Measures of Dispersion

**Objectives**

❶ Determine the Range of a Variable from Raw Data

❷ Determine the Standard Deviation of a Variable from Raw Data

❸ Determine the Variance of a Variable from Raw Data

❹ Use the Empirical Rule to Describe Data That Are Bell-Shaped

INTRODUCTION, PAGE 1

Measures of central tendency describe the typical value of a variable. We also want to know the amount of dispersion (or spread) in the variable. Dispersion is the degree to which the data are spread out.

INTRODUCTION, PAGE 2

| **Example 1** | *Comparing Two Sets of Data* |

The data tables represent the IQ scores of a random sample of 100 students from two different universities.
For each university, compute the mean IQ score and draw a histogram, using a lower class limit of 55 for the first class and a class width of 15. Comment on the results.

## Objective 1: Determine the Range of a Variable from Raw Data

OBJECTIVE 1, PAGE 1

1) What is the range of a variable?

OBJECTIVE 1, PAGE 2

| **Example 2** | **Computing the Range of a Set of Data** |

The data in the table represent the first exam scores of 10 students enrolled in Introductory Statistics. Compute the range.

| Student | Score |
|---|---|
| 1. Michelle | 82 |
| 2. Ryanne | 77 |
| 3. Bilal | 90 |
| 4. Pam | 71 |
| 5. Jennifer | 62 |
| 6. Dave | 68 |
| 7. Joel | 74 |
| 8. Sam | 84 |
| 9. Justine | 94 |
| 10. Juan | 88 |

## Objective 2: Determine the Standard Deviation of a Variable from Raw Data

OBJECTIVE 2, PAGE 1

2) Explain how to compute the population standard deviation $\sigma$ and list its formula.

OBJECTIVE 2, PAGE 2

---

**Example 3**     *Computing a Population Standard Deviation*

Compute the population standard deviation of the test scores in Table 6.

**Table 6**

| Student | Score |
|---|---|
| 1. Michelle | 82 |
| 2. Ryanne | 77 |
| 3. Bilal | 90 |
| 4. Pam | 71 |
| 5. Jennifer | 62 |
| 6. Dave | 68 |
| 7. Joel | 74 |
| 8. Sam | 84 |
| 9. Justine | 94 |
| 10. Juan | 88 |

---

OBJECTIVE 2, PAGE 5

3) If a data set has many values that are "far" from the mean, how is the standard deviation affected?

OBJECTIVE 2, PAGE 6

4) Explain how to compute the sample standard deviation $s$ and list its formula.

OBJECTIVE 2, PAGE 7

5) What do we call the expression $n-1$?

<u>OBJECTIVE 2, PAGE 8</u>

| **Example 4** | ***Computing a Sample Standard Deviation*** |

In a previous lesson we obtained a simple random sample of exam scores and computed a sample mean of 73.75. Compute the sample standard deviation of the sample of test scores for that data.

<u>OBJECTIVE 2, PAGE 10</u>

*Answer the following after you watch the video.*
6) Is standard deviation resistant? Why or why not?

<u>OBJECTIVE 2, PAGE 11</u>

7) When comparing two populations, what does a larger standard deviation imply about dispersion?

<u>OBJECTIVE 2, PAGE 14</u>

| **Example 5** | ***Comparing the Standard Deviations of Two Sets of Data*** |

The data tables represent the IQ scores of a random sample of 100 students from two different universities.
Use the standard deviation to determine whether University A or University B has more dispersion in the IQ scores of its students.

OBJECTIVE 2, PAGE 17

*Answer the following after using the applet in Activity 1: Standard Deviation as a Measure of Spread.*

8) Compare the dispersion of the observations in Part A with the observations in Part B. Which set of data is more spread out?

9) In Part D, how does adding a point near 10 affect the standard deviation? How is the standard deviation affected when that point is moved near 25? What does this suggest?

OBJECTIVE 2, PAGE 18

*Watch the video to reinforce the ideas from Activity 1: Standard Deviation as a Measure of Spread.*

## *Objective 3: Determine the Variance of a Variable from Raw Data*

OBJECTIVE 3, PAGE 1

10) Define variance.

OBJECTIVE 3, PAGE 2

**Example 6** ***Determining the Variance of a Variable for a Population and a Sample***

In previous examples, we considered population data of exam scores in a statistics class. For this data, we computed a population mean of $\mu = 79$ points and a population standard deviation of $\sigma = 9.8$ points. Then, we obtained a simple random sample of exam scores. For this data, we computed a sample mean of $\bar{x} = 73.75$ points and a sample standard deviation of $s = 11.3$ points. Use the population standard deviation exam score and the sample standard deviation exam score to determine the population and sample variance of scores on the statistics exam.

<u>OBJECTIVE 3, PAGE 3</u>

*Answer the following after you watch the video.*

11) Using a rounded value of the standard deviation to obtain the variance results in a round-off error. How should you deal with this issue?

<u>OBJECTIVE 3, PAGE 5</u>

Whenever a statistic consistently underestimates a parameter, it is said to be **biased**. To obtain an unbiased estimate of the population variance, divide the sum of the squared deviations about the sample mean by $n-1$.

## *Objective 4: Use the Empirical Rule to Describe Data That Are Bell-Shaped*

<u>OBJECTIVE 4, PAGE 1</u>

12) According to the Empirical Rule, if a distribution is roughly bell shaped, then approximately what percent of the data will lie within 1 standard deviation of the mean? What percent of the data will lie within 2 standard deviations of the mean? What percent of the data will lie within 3 standard deviations of the mean?

<u>OBJECTIVE 4, PAGE 2</u>

13) Sketch the third part of Figure 5.

**Example 7**    *Using the Empirical Rule*

Table 9 represents the IQs of a random sample of 100 students at a university.
A) Determine the percentage of students who have IQ scores within 3 standard deviations of the mean according to the Empirical Rule.
B) Determine the percentage of students who have IQ scores between 67.8 and 132.2 according to the Empirical Rule.
C) Determine the actual percentage of students who have IQ scores between 67.8 and 132.2.
D) According to the Empirical Rule, what percentage of students have IQ scores between 116.1 and 148.3?

**Table 9**

| 73  | 103 | 91   | 93  | 136 | 108 | 92  | 104 | 90  | 78  |
|-----|-----|------|-----|-----|-----|-----|-----|-----|-----|
| 108 | 93  | 91   | 78  | 81  | 130 | 82  | 86  | 111 | 93  |
| 102 | 111 | 125  | 107 | 80  | 90  | 122 | 101 | 82  | 115 |
| 103 | 110 | 84   | 115 | 85  | 83  | 131 | 90  | 103 | 106 |
| 71  | 69  | 97   | 130 | 91  | 62  | 85  | 94  | 110 | 85  |
| 102 | 109 | 105  | 97  | 104 | 94  | 92  | 83  | 94  | 114 |
| 107 | 94  | 1121 | 113 | 115 | 106 | 97  | 106 | 85  | 99  |
| 102 | 109 | 76   | 94  | 103 | 112 | 107 | 101 | 91  | 107 |
| 107 | 110 | 106  | 103 | 93  | 110 | 125 | 101 | 91  | 119 |
| 118 | 85  | 127  | 141 | 129 | 60  | 115 | 80  | 111 | 79  |

## Section 3.3
## Measures of Central Tendency and Dispersion from Grouped Data

**Objectives**

❶ Approximate the Mean of a Variable from Grouped Data

❷ Compute the Weighted Mean

❸ Approximate the Standard Deviation from a Frequency Distribution

---

*Objective 1: Approximate the Mean of a Variable from Grouped Data*

OBJECTIVE 1, PAGE 1

1) Explain how to find the class midpoint.

2) List the formulas for approximating the population mean and sample mean from a frequency distribution.

OBJECTIVE 1, PAGE 2

| Example 1 | *Approximating the Mean for Continuous Quantitative Data from a Frequency Distribution* |
|---|---|

The frequency distribution in Table 10 represents the five-year rate of return of a random sample of 40 large-blend mutual funds. Approximate the mean five-year rate of return.

**Table 10**

| Class (5-year rate of return) | Frequency |
|---|---|
| 8-8.99 | 2 |
| 9-9.99 | 2 |
| 10-10.99 | 4 |
| 11-11.99 | 1 |
| 12-12.99 | 6 |
| 13-13.99 | 13 |
| 14-14.99 | 7 |
| 15-15.99 | 3 |
| 16-16.99 | 1 |
| 17-17.99 | 0 |
| 18-18.99 | 0 |
| 19-19.99 | 1 |

<u>OBJECTIVE 1, PAGE 2 (CONTINUED)</u>

---

## Objective 2: Compute the Weighted Mean

<u>OBJECTIVE 2, PAGE 1</u>

3) When data values have different importance, or weights, associated with them, we compute the weighted mean. Explain how to compute the weighted mean and list its formula.

<u>OBJECTIVE 2, PAGE 2</u>

**Example 2**       *Computing the Weighted Mean*

Marissa just completed her first semester in college. She earned an A in her 4-hour statistics course, a B in her 3-hour sociology course, an A in her 3-hour psychology course, a C in her 5-hour computer programming course, and an A in her 1-hour drama course. Determine Marissa's grade point average.

## Objective 3: Approximate the Standard Deviation from a Frequency Distribution

<u>OBJECTIVE 3, PAGE 1</u>

4) List the formulas for approximating the population standard deviation and sample standard deviation of a variable from a frequency distribution.

<u>OBJECTIVE 3, PAGE 2</u>

| **Example 3** | ***Approximating the Standard Deviation from a Frequency Distribution*** |
| --- | --- |

The frequency distribution in Table 11 represents the five-year rate of return of a random sample of 40 large-blend mutual funds. Approximate the standard deviation five-year rate of return.

**Table 11**

| Class (5-year rate of return) | Frequency |
| --- | --- |
| 8-8.99 | 2 |
| 9-9.99 | 2 |
| 10-10.99 | 4 |
| 11-11.99 | 1 |
| 12-12.99 | 6 |
| 13-13.99 | 13 |
| 14-14.99 | 7 |
| 15-15.99 | 3 |
| 16-16.99 | 1 |
| 17-17.99 | 0 |
| 18-18.99 | 0 |
| 19-19.99 | 1 |

## Section 3.4
## Measures of Position

**Objective**

&#10112; Determine and Interpret *z*-Scores

&#10113; Interpret Percentiles

&#10114; Determine and Interpret Quartiles

&#10115; Determine and Interpret the Interquartile Range

&#10116; Check a Set of Data for Outliers

---

*Objective 1: Determine and Interpret z-Scores*

<u>OBJECTIVE 1, PAGE 1</u>

1) What does a *z*-score represent?

2) Explain how to find a *z*-score and list the formulas for computing a population *z*-score and a sample *z*-score.

3) What does a positive *z*-score for a data value indicate? What does a negative z-score indicate?

4) What does a *z*-score measure?

OBJECTIVE 1, PAGE 1 (CONTINUED)

5) How are *z*-scores rounded?

OBJECTIVE 1, PAGE 2

| Example 1 | *Determine and Interpret z-Scores* |

Determine whether the Boston Red Sox or the Colorado Rockies had a relatively better run-producing season. The Red Sox scored 878 runs and play in the American League, where the mean number of runs scored was $\mu = 731.3$ and the standard deviation was $\sigma = 54.9$ runs. The Rockies scored 845 runs and play in the National League, where the mean number of runs scored was $\mu = 718.3$ and the standard deviation was $\sigma = 61.7$ runs.

OBJECTIVE 1, PAGE 5

With negative *z*-scores, we need to be careful when deciding the better outcome. For example, when comparing finishing times for a marathon the lower score is better because it is more standard deviations below the mean.

## *Objective 2: Interpret Percentiles*

OBJECTIVE 2, PAGE 1

6) What does the *k*th percentile represent?

OBJECTIVE 2, PAGE 2

| Example 2 | *Interpreting a Percentile* |

Jennifer just received the results of her SAT exam. Her math score of 600 is at the 74th percentile. Interpret this result.

---

## Objective 3: Determine and Interpret Quartiles

OBJECTIVE 3, PAGE 1
7) Define the first, second, and third quartiles.

OBJECTIVE 3, PAGE 2
8) List the three steps for finding quartiles.

OBJECTIVE 3, PAGE 3

| Example 3 | Finding and Interpreting Quartiles |
|---|---|

The Highway Loss Data Institute routinely collects data on collision coverage claims. Collision coverage insures against physical damage to an insured individual's vehicle. Table 12 represents a random sample of 18 collision coverage claims based on data obtained from the Highway Loss Data Institute for 2007 models. Find and interpret the first, second, and third quartiles for collision coverage claims.

**Table 12**

| | | |
|---|---|---|
| $6751 | $9908 | $3461 |
| $2336 | $21,147 | $2332 |
| $189 | $1185 | $370 |
| $1414 | $4668 | $1953 |
| $10,034 | $735 | $802 |
| $618 | $180 | $1657 |

## Objective 4: Determine and Interpret the Interquartile Range

OBJECTIVE 4, PAGE 1

9) Which measure of dispersion is resistant?

10) Define the interquartile range, IQR.

OBJECTIVE 4, PAGE 2

| Example 4 | *Finding and Interpreting the Interquartile Range* |

Determine and interpret the interquartile range of the collision claim data from Table 12 in Example 3.

**Table 12**

| | | |
|---|---|---|
| $6751 | $9908 | $3461 |
| $2336 | $21,147 | $2332 |
| $189 | $1185 | $370 |
| $1414 | $4668 | $1953 |
| $10,034 | $735 | $802 |
| $618 | $180 | $1657 |

OBJECTIVE 4, PAGE 4

11) If the shape of a distribution is symmetric, which measure of central tendency and which measure of dispersion should be reported?

12) If the shape of a distribution is skewed left or skewed right, which measure of central tendency and which measure of dispersion should be reported? Why?

**83**

## *Objective 5: Check a Set of Data for Outliers*

<u>OBJECTIVE 5, PAGE 1</u>
13) What is an outlier?

<u>OBJECTIVE 5, PAGE 2</u>
14) List the four steps for checking for outliers by using quartiles.

<u>OBJECTIVE 5, PAGE 3</u>

**Example 5**    *Checking for Outliers*

Check the data in Table 12 on collision coverage claims for outliers.

**Table 12**

| | | |
|---|---|---|
| $6751 | $9908 | $3461 |
| $2336 | $21,147 | $2332 |
| $189 | $1185 | $370 |
| $1414 | $4668 | $1953 |
| $10,034 | $735 | $802 |
| $618 | $180 | $1657 |

## Section 3.5
## The Five-Number Summary and Boxplots

**Objectives**

❶ Determine the Five-Number Summary

❷ Draw and Interpret Boxplots

---

### Objective 1: Determine the Five-Number Summary

OBJECTIVE 1, PAGE 1

1) What values does the five-number summary consist of?

OBJECTIVE 1, PAGE 2

| **Example 1** | ***Obtaining the Five-Number Summary*** |

Table 13 shows the finishing times (in minutes) of the men in the 60- to 64-year-old age group in a 5-kilometer race. Determine the five-number summary of the data.

**Table 13**

| 19.95 | 23.25 | 23.32 | 25.55 | 25.83 | 26.28 | 42.47 |
| 28.58 | 28.72 | 30.18 | 30.35 | 30.95 | 32.13 | 49.17 |
| 33.23 | 33.53 | 36.68 | 37.05 | 37.43 | 41.42 | 54.63 |

Data from Laura Gillogly, student at Joliet Junior College

## Objective 2: Draw and Interpret Boxplots

<u>OBJECTIVE 2, PAGE 1</u>
2) List the five steps for drawing a boxplot.

<u>OBJECTIVE 2, PAGE 2</u>

| Example 2 | *Constructing a Boxplot* |

Use the results of Example 1 to construct a boxplot of the finishing times of the men in the 60- to 64-year-old age group.
(The five-number summary is: 19.95, 26.06, 30.95, 37.24, 54.63.)

<u>OBJECTIVE 2, PAGE 4</u>
3) If the right whisker of a boxplot is longer than the left whisker and the median is left of the center of the box, what is the most likely shape of the distribution?

OBJECTIVE 2, PAGE 5

When describing the shape of a distribution from a boxplot, be sure to justify your conclusion. Possible areas to discuss:

- Compare the length of the left whisker to the length of the right whisker
- The position of the median in the box
- Compare the distance between the median and the first quartile to the distance between the median and the third quartile
- Compare the distance between the median and the minimum value to the distance between the median and the maximum value

OBJECTIVE 2, PAGE 10

---

**Example 3**    *Comparing Two Distributions Using Boxplots*

Table 14 shows the red blood cell mass (in millimeters) for 14 rats sent into space (flight group) and for 14 rats that were not sent into space (control group). Construct side-by-side boxplots for red blood cell mass for the flight group and control group. Does it appear that space flight affects the rats' red blood cell mass?

**Table 14**

| Flight | | | | Control | | | |
|--------|------|------|------|---------|------|------|------|
| 7.43 | 7.21 | 8.59 | 8.64 | 8.65 | 6.99 | 8.40 | 9.66 |
| 9.79 | 6.85 | 6.87 | 7.89 | 7.62 | 7.44 | 8.55 | 8.70 |
| 9.30 | 8.03 | 7.00 | 8.80 | 7.33 | 8.58 | 9.88 | 9.94 |
| 6.39 | 7.54 | | | 7.14 | 9.14 | | |

Data from NASA Life Sciences Data Archive

---

# Chapter 4 – Describing the Relation between Two Variables

## OUTLINE

**Putting It Together**

So far, we have examined data in which a single variable was measured for each individual in the study (univariate data), such as the 5-year rate of return (the variable) for various mutual funds (the individuals). We found both graphical and numerical descriptive measures for the variable.

Now, we discuss graphical and numerical methods for describing **bivariate data**, data in which two variables are measured on an individual. For example, we might want to know whether the amount of cola consumed per week is related to one's bone density. The individuals would be the people in the study nd the two variables would be the amount of cola consumed weekly and bone density. In this study, both variables are quantitative.

Suppose we want to know whether level of education is related to one's employment status (employed or unemployed). Here, both variables are qualitative.

Situations may also occur in which one variable is quantitative and the other is qualitative. We have already presented a technique for describing this situation. Look back at Example 3 in Section 3.5 where we considered whether space flight affected red blood cell mass. In this case, space flight is qualitative (rat sent to space or not) and red blood cell mass is quantitative.

## Section 4.1
## Scatter Diagrams and Correlation

**Objectives**

❶ Draw and Interpret Scatter Diagrams

❷ Describe the Properties of the Linear Correlation Coefficient

❸ Compute and Interpret the Linear Correlation Coefficient

❹ Determine Whether a Linear Relation Exists between Two Variables

❺ Explain the Difference between Correlation and Causation

---

*Objective 1: Draw and Interpret Scatter Diagrams*

INTRODUCTION, PAGE 2
1) Define bivariate data.

2) Define response variable and explanatory variable.

OBJECTIVE 1, PAGE 1
3) What is a scatter diagram? How is it created?

**Example 1**     *Drawing a Scatter Diagram*

A golf pro wants to investigate the relation between the club-head speed of a golf club (measured in miles per hour) and the distance (in yards) the ball will travel. He realizes that other variables besides club-head speed also determine the distance a ball will travel (such as club type, ball type, golfer, and weather conditions). To eliminate the variability due to these variables, the pro uses a single model of club and ball, one golfer, and a clear, 70-degree day with no wind. The pro records the club-head speed, measures the distance the ball travels, and collects the data in Table 1. Draw a scatter diagram of the data.

**Table 1**

| Club-Head Speed (mph) | Distance (yards) |
|---|---|
| 100 | 257 |
| 102 | 264 |
| 103 | 274 |
| 101 | 266 |
| 105 | 277 |
| 100 | 263 |
| 99 | 258 |
| 105 | 275 |

Data from Paul Stephenson, student at Joliet Junior College

4) Explain how to determine which variable is the explanatory variable and which variable is the response variable.

OBJECTIVE 1, PAGE 6

5) Sketch a scatterplot that shows a nonlinear relation between an explanatory variable and a response variable.

6) Sketch a scatterplot that shows no relation between an explanatory variable and a response variable.

OBJECTIVE 1, PAGE 7

7) Explain how to determine if two variables are positively associated.

8) Explain how to determine if two variables are negatively associated.

## *Objective 2: Describe the Properties of the Linear Correlation Coefficient*

<u>OBJECTIVE 2, PAGE 1</u>

It is dangerous to use only a scatter diagram to determine if two variables are linearly related. Just as we can manipulate the scale of graphs of univariate data, we can also manipulate the scale of graphs of bivariate data, possibly resulting in incorrect conclusions.

<u>OBJECTIVE 2, PAGE 2</u>

9) Define the linear correlation coefficient and state the formula for the sample linear correlation coefficient.

<u>OBJECTIVE 2, PAGE 3</u>

*Answer the following after watching the video.*

10) In the formula for the linear correlation coefficient, notice that the numerator is the sum of the products of $z$-scores for the explanatory ($x$) and response ($y$) variables. If the linear correlation coefficient is positive, that means that the sum of the $z$-scores for $x$ and $y$ must be positive. How does that occur?

Chapter 4: Describing the Relation between Two Variables

OBJECTIVE 2, PAGE 4

*Answer the following as you work through Activity 1: Properties of the Correlation Coefficient.*
11) When the points are aligned in a straight line with positive slope, what is the value of the linear correlation coefficient?

12) When the points are aligned in a straight line with negative slope, what is the value of the linear correlation coefficient?

13) Does a correlation coefficient close to 0 imply that there is no relation? Why or why not?

14) Is the linear correlation coefficient resistant?

OBJECTIVE 2, PAGE 5

*Watch the video for a summary of the ideas from Activity 1.*

OBJECTIVE 2, PAGE 6
**Note: Properties of the Linear Correlation Coefficient**
The linear correlation coefficient is always between $-1$ and 1, inclusive. That is, $-1 \leq r \leq 1$.

If $r = +1$, then a perfect positive linear relation exists between the two variables.

If $r = -1$, then a perfect negative linear relation exists between the two variables.

The closer $r$ is to $+1$, the stronger is the evidence of positive association between the two variables.

The closer $r$ is to $-1$, the stronger is the evidence of negative association between the two variables.

If $r$ is close to 0, then little or no evidence exists of a linear relation between the two variables. So $r$ close to 0 does not imply no relation, just no linear relation.

The linear correlation coefficient is a unitless measure of association. So the unit of measure for $x$ and $y$ plays no role in the interpretation of $r$.

The correlation coefficient is not resistant. Therefore, an observation that does not follow the overall pattern of the data could affect the value of the linear correlation coefficient.

**94**
Copyright © 2019 Pearson Education, Inc.

## *Objective 3: Compute and Interpret the Linear Correlation Coefficient*

OBJECTIVE 3, PAGE 1
Linear correlation coefficients are typically found using technology. Example 2 shows how it would be computed by hand to help show how it measures the strength of a linear relation.

OBJECTIVE 3, PAGE 2

**Example 2**     *Computing the Correlation Coefficient by Hand*

For the data shown in Table 2, compute the linear correlation coefficient. A scatter diagram of the data is shown in Figure 5. The dashed lines on the scatter diagram represent the mean of x and y.

**Table 2**

| $x$ | $y$ |
|-----|-----|
| 1 | 18 |
| 3 | 13 |
| 3 | 9 |
| 6 | 6 |
| 7 | 4 |

**Figure 5**

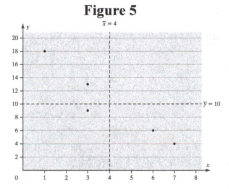

95

<u>OBJECTIVE 3, PAGE 3</u>

In the scatter diagram, notice that below-average values of $x$ are associated with above-average values of $y$ and above-average values of $x$ are associated with below-average values of $y$. This helps to explain why the linear correlation coefficient is negative.

<u>OBJECTIVE 3, PAGE 5</u>

---

**Example 3**     *Determining the Linear Correlation Coefficient Using Technology*

Use a statistical spreadsheet or a graphing calculator with advanced statistical features to determine the linear correlation coefficient between club-head speed and distance from the data in Table 1. Interpret the linear correlation coefficient.

**Table 1**

| Club-Head Speed (mph) | Distance (yards) |
|:---:|:---:|
| 100 | 257 |
| 102 | 264 |
| 103 | 274 |
| 101 | 266 |
| 105 | 277 |
| 100 | 263 |
| 99 | 258 |
| 105 | 275 |

Data from Paul Stephenson, student at Joliet Junior College

---

*Objective 4: Determine Whether a Linear Relation Exists between Two Variables*

<u>OBJECTIVE 4, PAGE 1</u>

15) List the three steps for testing for a linear relation.

---

**Example 4**    *Does a Linear Relation Exist?*

We have been analyzing the association between club-head speed and distance the ball travels. A scatter diagram suggests a positive association between the variables. The linear correlation coefficient of 0.939 also suggests a positive association between the variables. Use these results to determine whether a linear relation exists between club-head speed and distance.

---

## Objective 5: Explain the Difference between Correlation and Causation

16) If data are obtained from an experiment, can we claim a causal relationship between the explanatory and response variables? How about if the data are obtained from an observational study?

17) Is there another way two variables can be correlated without a causal relationship existing?

**Example 5**     *Lurking Variables in a Bone Mineral Density Study*

Because cola tends to replace healthier beverages and cola contains caffeine and phosphoric acid, researchers Katherine L. Tucker and associates wanted to know if cola consumption is associated with lower bone mineral density in women. Table 4 lists the typical number of cans of cola consumed in a week and the bone mineral density for a sample of 15 women. The data were collected through a prospective cohort study.

Figure 7 shows the scatter diagram of the data. The correlation between number of colas per week and bone mineral density is −0.806. The critical value for correlation with $n = 15$ from Table II is 0.514. Because $|-0.806| > 0.514$, we conclude that a negative linear relation exists between number of colas consumed and bone mineral density. Can the authors conclude that an increase in the number of colas consumed causes a decrease in bone mineral density? Identify some lurking variables in the study.

# Section 4.2
# Least-Squares Regression

**Objectives**

❶ Find the Least-Squares Regression Line and Use the Line to Make Predictions

❷ Interpret the Slope and the *y*-Intercept of the Least-Squares Regression Line

❸ Compute the Sum of Squared Residuals

INTRODUCTION, PAGE 2

**Example 1**    *Finding an Equation That Describes Linearly Related Data*

The data in Table 1 represent the club-head speed and the distance a golf ball travels for eight swings of the club.

**Table 1**

| Club-Head Speed (mph) | Distance (yards) | (*x*, *y*) |
|---|---|---|
| 100 | 257 | (100, 257) |
| 102 | 264 | (102, 264) |
| 103 | 274 | (103, 274) |
| 101 | 266 | (101, 266) |
| 105 | 277 | (105, 277) |
| 100 | 263 | (100, 263) |
| 99 | 258 | (99, 258) |
| 105 | 275 | (105, 275) |

Data from Paul Stephenson, student at Joliet Junior College

A) Find a linear equation that relates club-head speed x (the explanatory variable) and distance y (the response variable) by selecting two points and finding the equation of the line containing the points.

B) Graph the line on the scatter diagram.

C) Use the equation to predict the distance a golf ball will travel if the club-head speed is 104 miles per hour.

## *Objective 1: Find the Least-Squares Regression Line and Use the Line to Make Predictions*

<u>OBJECTIVE 1, PAGE 1</u>

1) What is the residual for an observation?

<u>OBJECTIVE 1, PAGE 3</u>

2) What is the least-squares regression line?

<u>OBJECTIVE 1, PAGE 4</u>

*Answer the following as you work through Activity 1: What is Least Squares?*

3) As your line gets closer and closer to the regression line, what happens to the SSE?

<u>OBJECTIVE 1, PAGE 5</u>

4) List the formulas associated with the least-squares regression line.

OBJECTIVE 1, PAGE 6

**Key Ideas about the Least-Squares Regression Line**

- The least-squares regression line, $\hat{y} = b_1 x + b_0$, always contains the point $(\bar{x}, \bar{y})$.

- Because $s_y$ and $s_x$ must both be positive, the sign of the linear correlation coefficient, $r$, and the sign of the slope of the least-squares regression line, $b_1$, are the same.

- The predicted value of $y$, $\hat{y}$, has an interesting interpretation. It is an estimate of the mean value of the response variable for any value of the explanatory variable.

OBJECTIVE 1, PAGE 8

---

**Example 2**  *Finding the Least-Squares Regression Line by Hand*

Find the least-squares regression line for the data in Table 2 from Section 4.1. For convenience, we also provide a scatter diagram of the data.

**Table 2**

| $x$ | $y$ |
|-----|-----|
| 1   | 18  |
| 3   | 13  |
| 3   | 9   |
| 6   | 6   |
| 7   | 4   |

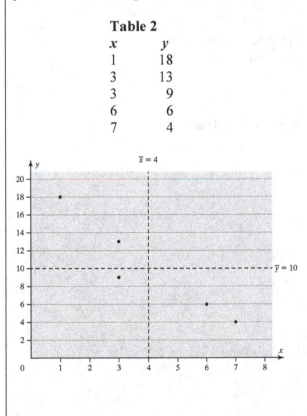

---

Chapter 4: Describing the Relation between Two Variables

OBJECTIVE 1, PAGE 9
Throughout the course, we agree to round the slope and $y$-intercept to four decimal places.
In Example 2 we found a least-squares regression line to see the role that the correlation coefficient, $r$, the standard deviation of $y$, and the standard deviation of $x$ play in finding the slope.
In practice, the least-squares regression line is found using technology.

OBJECTIVE 1, PAGE 10

| Example 3 | Finding the Least-Squares Regression Line Using Technology |
| --- | --- |

Use the golf data in Table 1.

**Table 1**

| Club-Head Speed (mph) | Distance (yards) |
| --- | --- |
| 100 | 257 |
| 102 | 264 |
| 103 | 274 |
| 101 | 266 |
| 105 | 277 |
| 100 | 263 |
| 99 | 258 |
| 105 | 275 |

Data from Paul Stephenson, student at Joliet Junior College

A) Find the equation of the least-squares regression line using technology.

B) Draw the least-squares regression line on the scatter diagram of the data.

C) Predict the distance a golf ball will travel when hit with a club-head speed of 103 miles per hour (mph).

D) Determine the residual for the predicted value found in part (c). Is the distance above or below average among all balls hit with a swing speed of 103 mph?

## Objective 2: Interpret the Slope and the y-Intercept of the Least-Squares Regression Line

OBJECTIVE 2, PAGE 1

5) The slope of the least-squares regression line from Example 3 is 3.1661 yards per mph. List two interpretations of the slope that are acceptable.

OBJECTIVE 2, PAGE 2

6) Before interpreting the *y*-intercept, what two questions must be asked?

7) What does it mean to say that we should not use the regression model to make predictions outside the scope of the model?

OBJECTIVE 2, PAGE 3

8) When the correlation coefficient indicates no linear relation between the explanatory and response variables and the scatter diagram indicates no relation between the variables, how do we find a predicted value for the response variable?

## Objective 3: Compute the Sum of Squared Residuals

OBJECTIVE 3, PAGE 1

9) What does the least-squares regression line minimize?

---

**Example 4**    *Computing the Sum of Squared Residuals*

Compare the sum of squared residuals for the linear models in Examples 1 and 3.

---

OBJECTIVE 3, PAGE 3

The total area of the red squares represent the sum of the squared residuals for the least-squares regression line. Any other line will produce a larger sum of the squared residuals, such as the sum of the areas of the green squares.

# Section 4.3
# Diagnostics on the Least-Squares Regression Line

**Objectives**

❶ Compute and Interpret the Coefficient of Determination

❷ Perform Residual Analysis on a Regression Model

❸ Identify Influential Observations

---

## Objective 1: Compute and Interpret the Coefficient of Determination

OBJECTIVE 1, PAGE 1

1) Define the coefficient of determination, $R^2$.

2) The coefficient of determination is a number between 0 and 1, inclusive. That is, $0 \le R^2 \le 1$. What does it mean if $R^2 = 0$? What does it mean if $R^2 = 1$?

OBJECTIVE 1, PAGE 2

The differences between the observed value $(y)$, the mean value $(\bar{y})$ of the response variable, and the predicted value $(\hat{y})$ are called deviations.

OBJECTIVE 1, PAGE 4

3) Explain the meaning of total deviation, explained deviation, and unexplained deviation.

4) The closer the observed $y$'s are to the regression line (the predicted $y$'s), how is $R^2$ affected?

5) Explain how to find the coefficient of determination, $R^2$, for the least-squares regression model $\hat{y} = b_1 x + b_0$.

**Example 1**     *Determining and Interpreting the Coefficient of Determination*

Determine and interpret the coefficient of determination, R2, for the club-head speed versus distance data shown in Table 1.

**Table 1**

| Club-Head Speed (mph) | Distance (yards) |
|:---:|:---:|
| 100 | 257 |
| 102 | 264 |
| 103 | 274 |
| 101 | 266 |
| 105 | 277 |
| 100 | 263 |
| 99 | 258 |
| 105 | 275 |

Data from Paul Stephenson, student at Joliet Junior College

*Fill in the table as you work through Activity 1: Understanding the Coefficient of Determination.*

6)

| Data Set | Coefficient of Determination, $R^2$ | Interpretation |
|---|---|---|
| A | _____ | ____ of the variability in y is explained by the least-squares regression line. |
| B | _____ | ____ of the variability in y is explained by the least-squares regression line. |
| C | _____ | ____ of the variability in y is explained by the least-squares regression line. |

***Objective 2: Perform Residual Analysis on a Regression Model***

OBJECTIVE 2, PAGE 1

7) List the three purposes for which we analyze residuals.

OBJECTIVE 2, PAGE 2

*Answer the following after watching the video.*

8) What is a residual plot?

9) If a plot of the residuals against the explanatory variable shows a discernable pattern, what does this say about the explanatory and response variables?

OBJECTIVE 2, PAGE 3

*Answer the following after watching the video.*

10) Why is it important for the residuals to have constant error variance?

OBJECTIVE 2, PAGE 4

*Answer the following after watching the third video.*

11) List two ways that we can use residuals to identify outliers.

## Example 2  *Graphical Residual Analysis*

The data in Table 6 represent the club-head speed, the distance the golf ball traveled, and the residuals (Column 4) for eight swings of a golf club. Construct a residual plot and boxplot of the residuals and comment on the appropriateness of the least-squares regression model.

**Table 6**

| Club-Head Speed (mph) | Example 3 Distance(yd) | $\hat{y} = 3.1661x - 55.7966$ | Residual $y - \hat{y}$ |
|---|---|---|---|
| 100 | 257 | 260.8 | −3.8 |
| 102 | 264 | 267.1 | −3.1 |
| 103 | 274 | 270.3 | 3.7 |
| 101 | 266 | 264.0 | 2.0 |
| 105 | 277 | 276.6 | 0.4 |
| 100 | 263 | 260.8 | 2.2 |
| 99 | 258 | 257.6 | 0.4 |
| 105 | 275 | 276.6 | −1.6 |

## Objective 3: Identify Influential Observations

OBJECTIVE 3, PAGE 1

12) What is an influential observation?

OBJECTIVE 3, PAGE 2

📱*Answer the following after working through Activity 2: Influential Observations.*

13) For the point that had the greatest effect on the slope and *y*-intercept, how did its residual compare to the other points you added? How did its *x*-value compare to the other points you added?

OBJECTIVE 3, PAGE 5

14) What do we call the relative vertical position of an observation? What do we call the relative horizontal position of an observation?

15) List a combination of leverage and residual that indicates that an observation may be influential.

OBJECTIVE 3, PAGE 6

| **Example 3** | *Identifying Influential Observations* |
| --- | --- |

Suppose that our golf ball experiment calls for nine trials, but the player we are using hurts his wrist. Luckily, Bubba Watson (a professional golfer) is practicing on the same range and graciously agrees to participate in our experiment. His club-head speed is 120 miles per hour, and he hits the golf ball 305 yards. Is Bubba's shot an influential observation? For convenience, we present the data from Table 1.

**Table 1**

| Club-Head Speed (mph) | Distance (yards) |
| --- | --- |
| 100 | 257 |
| 102 | 264 |
| 103 | 274 |
| 101 | 266 |
| 105 | 277 |
| 100 | 263 |
| 99 | 258 |
| 105 | 275 |

Data from Paul Stephenson, student at Joliet Junior College

<u>OBJECTIVE 3, PAGE 6 (CONTINUED)</u>

<u>OBJECTIVE 3, PAGE 7</u>

As with outliers, influential observations should be removed only if there is justification to do so. When an influential observation occurs in a data set and its removal is not warranted, two possible courses of action are to (1) collect more data so that additional points near the influential observation are obtained or (2) use techniques that reduce the influence of the influential observation. (These techniques are beyond the scope of this text.)

## Section 4.4
## Contingency Tables and Association

**Objectives**

❶ Compute the Marginal Distribution of a Variable

❷ Use the Conditional Distribution to Identify Association Among Categorical Data

❸ Explain Simpson's Paradox

<u>INTRODUCTION, PAGE 1</u>

1) What is a contingency table?

---

*Objective 1: Compute the Marginal Distribution of a Variable*

<u>OBJECTIVE 1, PAGE 1</u>

2) What is a marginal distribution of a variable?

**Example 1**    *Determining Frequency Marginal Distributions*

The data (measured in thousands) in Table 8 represent the employment status and level of education of all civilian, noninstitutional residents (excludes inmates and individuals in the armed services) who are 25 years and older in May 2017. Find the frequency marginal distributions for employment status and level of education.

**Table 8**

| Employment Status | Level of Education | | | |
| --- | --- | --- | --- | --- |
| | Did Not Finish High School | High School Graduate | Some College | Bachelor's Degree or Higher |
| Employed | 9671 | 34,211 | 35,941 | 53,760 |
| Unemployed | 628 | 1697 | 1492 | 1278 |
| Not in the Labor Force | 12,564 | 26,406 | 19,346 | 19,524 |

*Data from Bureau of Labor Statistics*

**Example 2**     *Determining Relative Frequency Marginal Distributions*

Determine and interpret the relative frequency marginal distribution for level of education and employment status from the data in Table 9.

**Table 9**

| Employment Status | Did Not Finish High School | High School Graduate | Some College | Bachelor's Degree or Higher | Frequency Marginal Distribution for Employment Status |
|---|---|---|---|---|---|
| Employed | 9671 | 34,211 | 35,941 | 53,760 | 133,583 |
| Unemployed | 628 | 1697 | 1492 | 1278 | 5095 |
| Not in the Labor Force | 12,564 | 26,406 | 19,346 | 19,524 | 77,840 |
| Frequency Marginal Distribution for Level of Education | 22,863 | 62,314 | 56,779 | 74,562 | 216,518 |

## Objective 2: Use the Conditional Distribution to Identify Association Among Categorical Data

OBJECTIVE 2, PAGE 1
Marginal distributions allow us to see the distribution of either the row variable or the column variable, but we do not get a sense of association between the two variables from these tables.
To learn about any association that may exist, we need a different table.

OBJECTIVE 2, PAGE 2

**Example 3**    *Comparing Two Categories of a Variable*

Use the data in Table 9 to answer parts (A) through (D).
A) What proportion of those who did not finish high school is employed?
B) What proportion of those who are high school graduates is employed?
C) What proportion of those who finished some college is employed?
D) What proportion of those who have at least a Bachelor's degree is employed?

**Table 9**

| Employment Status | Did Not Finish High School | High School Graduate | Some College | Bachelor's Degree or Higher | Frequency Marginal Distribution for Employment Status |
|---|---|---|---|---|---|
| **Employed** | 9671 | 34,211 | 35,941 | 53,760 | 133,583 |
| **Unemployed** | 628 | 1697 | 1492 | 1278 | 5095 |
| **Not in the Labor Force** | 12,564 | 26,406 | 19,346 | 19,524 | 77,840 |
| **Frequency Marginal Distribution for Level of Education** | 22,863 | 62,314 | 56,779 | 74,562 | 216,518 |

OBJECTIVE 2, PAGE 3
3) What is a conditional distribution?

The variable we condition upon represents the explanatory variable in a conditional distribution, and the remaining variable becomes the response variable.

## Example 4    *Constructing a Conditional Distribution*

Find the conditional distribution of the response variable employment status by level of education, the explanatory variable, for the data in in Table 9. What is the association between level of education and employment status?

**Table 9**

| Employment Status | Did Not Finish High School | High School Graduate | Some College | Bachelor's Degree or Higher | Frequency Marginal Distribution for Employment Status |
|---|---|---|---|---|---|
| Employed | 9671 | 34,211 | 35,941 | 53,760 | 133,583 |
| Unemployed | 628 | 1697 | 1492 | 1278 | 5095 |
| Not in the Labor Force | 12,564 | 26,406 | 19,346 | 19,524 | 77,840 |
| Frequency Marginal Distribution for Level of Education | 22,863 | 62,314 | 56,779 | 74,562 | 216,518 |

**Example 5**     *Drawing a Bar Graph of a Conditional Distribution*

Using the results of Example 4, draw a bar graph that represents the conditional distribution of employment status by level of education.

The methods presented for identifying the association between two categorical variables are different from the methods used for measuring association between two quantitative variables. The measure of association is based on whether there are differences in the relative frequencies of the response variable (employment status) for the different categories of the explanatory variable (level of education). If differences exist, we might attribute these differences to the explanatory variable.

---

## Objective 3: Explain Simpson's Paradox

A lurking variable can cause two quantitative variables to be correlated even though they are unrelated. The same phenomenon exists when we explore the relation between two qualitative variables.

**Example 6**  *Gender Bias at the University of California, Berkeley*

The data in Table 12 show the admission status and gender of students who applied to the University of California, Berkeley. From the data in Table 12, the proportion of accepted applications is $\frac{1748}{4425} = 0.395$.

The proportion of accepted men is $\frac{1191}{2590} = 0.460$, and the proportion of accepted women is

$\frac{557}{1835} = 0.304$. On the basis of these proportions, a gender bias suit was brought against the university.

The university was shocked and claimed that program of study is a lurking variable that created the apparent association between admission status and gender. The university supplied Table 13 in its defense. Develop a conditional distribution by program of study and use it to defend the university's admission policies.

(Data from P. J. Bickel, E. A. Hammel, and J. W. O'Connell. "Sex Bias in Graduate Admissions: Data from Berkeley." Science 187(4175): 398-404, 1975.)

**Table 12**

|  | Accepted (A) | Not Accepted (NA) | Total |
|---|---|---|---|
| **Men** | 1191 | 1399 | **2590** |
| **Women** | 557 | 1278 | **1835** |
| **Total** | **1748** | **2677** | **4425** |

**Table 13**

| ADMISSION STATUS (ACCEPTED, A, OR NOT ACCEPTED, NA), FOR SIX PROGRAMS OF STUDY (A, B, C, D, E, F) BY GENDER | | | | | | | | | | | |
|---|---|---|---|---|---|---|---|---|---|---|---|
|  | **A** | | **B** | | **C** | | **D** | | **E** | | **F** | |
| **Men** | A | NA | A | NA | A | NA | A | NA | A | NA | A | NA |
|  | 511 | 314 | 353 | 207 | 120 | 205 | 138 | 279 | 53 | 138 | 16 | 256 |
| **Women** | A | NA | A | NA | A | NA | A | NA | A | NA | A | NA |
|  | 89 | 19 | 17 | 8 | 202 | 391 | 131 | 244 | 94 | 299 | 24 | 317 |

Simpson's Paradox describes a situation in which an association between two variables inverts or goes away when a third variable is introduced to the analysis.

# Chapter 5 – Probability

## Putting It Together

In Chapter 1, we learned the methods of collecting data. In Chapters 2 through 4, we learned how to summarize raw data using tables, graphs, and numbers. As far as the statistical process goes, we have discussed the collecting, organizing, and summarizing parts of the process.

Before we begin to analyze data, we introduce probability, which forms the basis of inferential statistics. Why? Well, we can think of the probability of an outcome as the likelihood of observing that outcome. If something has a high likelihood of happening, it has a high probability (close to 1). If something has a small chance of happening, it has a low probability (close to 0). For example, it is unlikely that we would roll five straight sixes when rolling a single die, so this result has a low probability. In fact, the probability of rolling five straight sixes is 0.0001286. If we were playing a game in which a player threw five sixes in a row with a single die, we would consider the player to be lucky (or a cheater) because it is such an unusual occurrence. Statisticians use probability in the same way. If something occurs that has a low probability, we investigate to find out "what's up."

## Section 5.1
## Probability Rules

**Objectives**

❶ Understand Random Processes and the Law of Large Numbers

❷ Apply the Rules of Probabilities

❸ Compute and Interpret Probabilities Using the Empirical Method

❹ Compute and Interpret Probabilities Using the Classical Method

❺ Recognize and Interpret Subjective Probabilities

---

*Objective 1: Understand Random Processes and the Law of Large Numbers*

<u>OBJECTIVE 1, PAGE 1</u>

*Answer the following while watching the video.*

1) Define simulation.

2) Define a random process.

The **short run** is a few repetitions of the simulation, while the **long run** is many repetitions of the simulation. In this video, there is a lot of variability in the proportion of heads observed in the short run, while in the long run the proportion of heads approaches 0.5.

<u>OBJECTIVE 1, PAGE 6</u>

3) Define probability.

4) State the Law of Large Numbers.

OBJECTIVE 1, PAGE 7

*Answer the following after finishing Activity 1: The Law of Large Numbers.*

5) After rolling the die 1000 times, is the behavior in the short run (fewer rolls of the die) the same as it was with the first 1000 runs? Based on your results, what is the probability of rolling a 4 with a 10-sided die?

OBJECTIVE 1, PAGE 9

*Answer the following after watching the video.*

6) Explain the meaning of the sentence, "In a random process, the trials are memoryless."

7) For a family whose first four children are girls, is the family more likely on the fifth child to have a boy?

OBJECTIVE 1, PAGE 11

8) In probability, what is an experiment?

OBJECTIVE 1, PAGE 13

| **Example 1** | ***Identifying Events and the Sample Space of a Probability Experiment*** |
| --- | --- |

A probability experiment consists of rolling a single six-sided fair die. A fair die is one in which each possible outcome is equally likely. For example, rolling a two is just as likely as rolling a five.

A) Identify the outcomes of the probability experiment.
B) Define the sample space.
C) Define the event E = "roll an even number."

## Objective 2: *Apply the Rules of Probabilities*

<u>OBJECTIVE 2, PAGE 1</u>

9) State Rules 1 and 2 of the rules of probabilities.

10) What is a probability model?

<u>OBJECTIVE 2, PAGE 2</u>

---

**Example 2**     *A Probability Model*

In a bag of peanut M&M milk chocolate candies, the colors of the candies can be brown, yellow, red, blue, orange, or green. Suppose that a candy is randomly selected from a bag. The table shows each color and the probability of drawing that color. Verify this is a probability model.

| Color | Probability |
|-------|-------------|
| Brown | 0.12 |
| Yellow | 0.15 |
| Red | 0.12 |
| Blue | 0.23 |
| Orange | 0.23 |
| Green | 0.15 |

---

<u>OBJECTIVE 2, PAGE 4</u>

- If an event is impossible, the probability of the event is 0.
- If an event is a certainty, the probability of the event is 1.
- The closer a probability is to 1, the more likely the event will occur.
- The closer a probability is to 0, the less likely the event will occur.
- For example, an event with probability 0.8 is more likely to occur than an event with probability 0.75.
- An event with probability 0.8 will occur about 80 times out of 100 repetitions of the experiment, whereas an event with probability 0.75 will occur about 75 times out of 100.

OBJECTIVE 2, PAGE 6

11) What is an unusual event? What cutoff points do statisticians typically use for identifying unusual events?

OBJECTIVE 2, PAGE 8

12) List the three methods in this section for determining the probability of an event.

## *Objective 3: Compute and Interpret Probabilities Using the Empirical Method*

OBJECTIVE 3, PAGE 1

13) Explain how to approximate probabilities using the empirical approach.

OBJECTIVE 3, PAGE 2

| Example 3 | *Using Relative Frequencies to Approximate Probabilities* |
|---|---|

An insurance agent currently insures 182 teenage drivers (ages 16 to 19). Last year, 24 of the teenagers had to file a claim on their auto policy. Based on these results, the probability that a teenager will file a claim on his or her auto policy in a given year is

---

**Example 4**     *Building a Probability Model from a Random Process*

Pass the Pigs™ is a Milton-Bradley game in which pigs are used as dice. Points are earned based on the way the pig lands. There are six possible outcomes when one pig is tossed. A class of 52 students rolled pigs 3939 times. The number of times each outcome occurred is recorded in the table.
(*Source:* www.members.tripod.com/~passpigs/prob.html)

| Outcome | Frequency |
|---|---|
| Side with no dot | 1344 |
| Side with dot | 1294 |
| Razorback | 767 |
| Trotter | 365 |
| Snouter | 137 |
| Leaning Jowler | 32 |

A) Use the results of the experiment to build a probability model for the way the pig lands.
B) Estimate the probability that a thrown pig lands on the "side with the dot".
C) Would it be unusual to throw a "Leaning Jowler"?

---

Surveys are probability experiments. Why? Each time a survey is conducted, a different random sample of individuals is selected. Therefore, the results of a survey are likely to be different each time the survey is conducted because different people are included.

---

## *Objective 4: Compute and Interpret Probabilities Using the Classical Method*

14) What requirement must be met in order to compute probabilities using the classical method?

<u>OBJECTIVE 4, PAGE 1 (CONTINUED)</u>

15) Explain how to compute probabilities using the classical method.

<u>OBJECTIVE 4, PAGE 2</u>

**Example 5**     *Computing Probabilities Using the Classical Approach*

A pair of fair dice is rolled. Fair die are die where each outcome is equally likely. The possible outcomes of this experiment are shown in Figure 1.

**Figure 1**

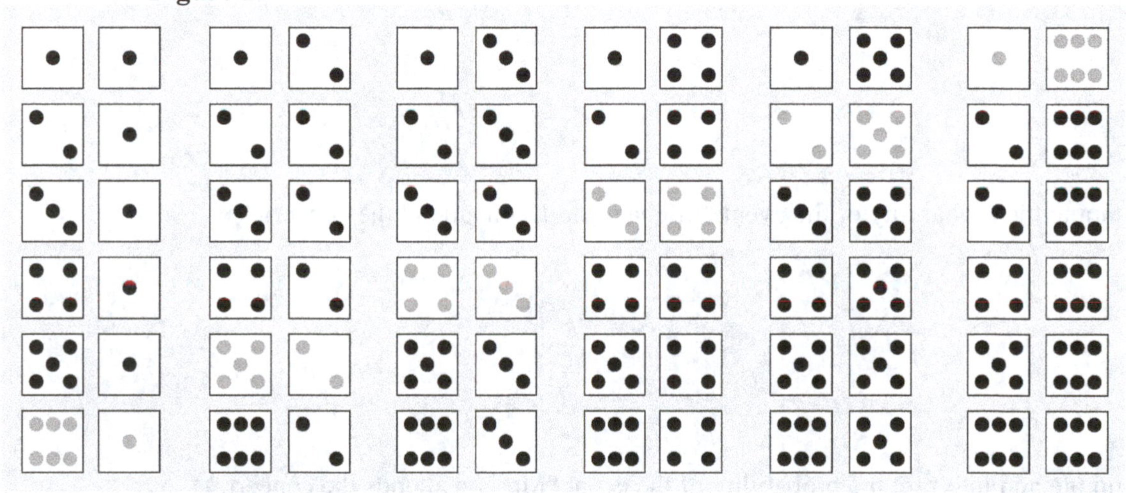

A) Compute the probability of rolling a seven.
B) Compute the probability of rolling "snake eyes"; that is, compute the probability of rolling a two.
C) Comment on the likelihood of rolling a seven versus rolling a two.

16) As the number of trials of an experiment increase, how does the empirical probability of an event occurring compare to the classical probability of that event occurring?

| Example 6 | *Computing Probabilities Using Equally Likely Outcomes* |
|---|---|

Sophia has three tickets to a concert, but Yolanda, Michael, Kevin, and Marissa all want to go to the concert with her. To be fair, Sophia wants to randomly select the two people who will go with her.

A) Determine the sample space of the experiment. In other words, list all possible simple random samples of size $n = 2$.

B) Compute the probability of the event "Michael and Kevin attend the concert."

C) Compute and interpret the probability of the event "Marissa attends the concert."

> **Example 7** *Comparing the Classical Method and Empirical Method*
>
> Suppose that a survey asked 500 families with three children to disclose the gender of their children and found that 180 of the families had two boys and one girl.
>
> A) Estimate the probability of having two boys and one girl in a three-child family, using the empirical method.
>
>
>
>
> B) Compute and interpret the probability of having two boys and one girl in a three-child family, using the classical method and assuming boys and girls are equally likely.

Empirical probabilities and classical probabilities often differ in value, but as the number of repetitions of a probability experiment increases, the empirical probability should get closer to the classical probability according to the Law of Large Numbers.

## Objective 5: Recognize and Interpret Subjective Probabilities

17) What is a subjective probability? Explain why subjective probabilities are used.

## Section 5.2
## The Addition Rule and Complements

**Objectives**

❶ Use the Addition Rule for Disjoint Events

❷ Use the General Addition Rule

❸ Compute the Probability of an Event Using the Complement Rule

---

### *Objective 1: Use the Addition Rule for Disjoint Events*

<u>OBJECTIVE 1, PAGE 1</u>

*Answer the following as you watch the video.*

1) What does it mean for two events to be disjoint?

2) In a Venn diagram, what does the rectangle represent? What does a circle represent?

3) How can you tell from a Venn diagram that two events are not disjoint?

4) For disjoint events $E$ and $F$, how is $P(E \text{ or } F)$ related to $P(E)$ and $P(F)$?

5) State the Addition Rule for Disjoint Events.

OBJECTIVE 1, PAGE 3

**Example 1**    *Benford's Law and the Addition Rule for Disjoint Events*

Our number system consists of the digits 0, 1, 2, 3, 4, 5, 6, 7, 8, and 9. Because we do not write numbers such as 12 as 012, the first significant digit in any number must be 1, 2, 3, 4, 5, 6, 7, 8, or 9. Although we may think that each digit appears with equal frequency so that each digit has a $\frac{1}{9}$ probability of being the first significant digit, this is not true. In 1881, Simon Newcomb discovered that first-digits do not occur with equal frequency. The physicist Frank Benford discovered the same result in 1938. After studying a great deal of data, he assigned probabilities of occurrence for each of the first digits, as shown in Table 2. The probability model is now known as Benford's Law and plays a major role in identifying fraudulent data on tax returns and accounting books.

**Table 2**

| Digit | 1 | 2 | 3 | 4 | 5 | 6 | 7 | 8 | 9 |
|-------|-------|-------|-------|-------|-------|-------|-------|-------|-------|
| Probability | 0.301 | 0.176 | 0.125 | 0.097 | 0.079 | 0.067 | 0.058 | 0.051 | 0.046 |

Data from The First Digit Phenomenon, T. P. Hill, American Scientist, July–August, 1998

A) Verify that Benford's Law is a probability model.
B) Use Benford's Law to determine the probability that a randomly selected first digit is 1 or 2.
C) Use Benford's Law to determine the probability that a randomly selected first digit is at least 6.

OBJECTIVE 1, PAGE 4

**Example 2**    *A Deck of Cards and the Addition Rule for Disjoint Events*

Suppose that a single card is selected from a standard 52-card deck, such as the one shown in Figure 3.

**Figure 3**

A) Compute the probability of the event $E$="drawing a king."
B) Compute the probability of the event $E$="drawing a king" or $F$="drawing a queen" or $G$="drawing a jack."

---

## Objective 2: Use the General Addition Rule

OBJECTIVE 2, PAGE 1

6) State the General Addition Rule.

7) Explain why we subtract $P(E \text{ and } F)$ when using the General Addition Rule.

OBJECTIVE 2, PAGE 3

**Example 3**   *Computing Probabilities for Events That Are Not Disjoint*

Suppose a single card is selected from a standard 52-card deck. Compute the probability of the event $E$="drawing a king" or $F$="drawing a diamond."

OBJECTIVE 2, PAGE 5

A table that relates two categories of data is called a **contingency table** (or **two-way table**).

The **row variable** is the variable that describes each row in the contingency table.

The **column variable** is the variable that describes each column in the contingency table.

OBJECTIVE 2, PAGE 6

**Example 4**     *Using the Addition Rule with Contingency Tables*

Use the data in Table 3 to answer parts (A) through (D).

**Table 3**

| Marital Status | Gender Males (in millions) | Females (in millions) |
|---|---|---|
| Never married | 44.1 | 39.0 |
| Married | 66.7 | 67.5 |
| Widowed | 3.5 | 11.4 |
| Divorced | 10.7 | 14.8 |

Data from U.S. Census Bureau, Current Population Reports

A) Determine the probability that a randomly selected U.S. resident 15 years and older is male.

B) Determine the probability that a randomly selected U.S. resident 15 years and older is widowed.

C) Determine the probability that a randomly selected U.S. resident 15 years and older is widowed or divorced.

D) Determine the probability that a randomly selected U.S. resident 15 years and older is male or widowed.

---

## Objective 3: Compute the Probability of an Event Using the Complement Rule

8) Define the complement of an event $E$.

9) State the Complement Rule.

| **Example 5** *Illustrating the Complement Rule* |
| --- |
| According to the American Veterinary Medical Association, 31.6% of American households own a dog. What is the probability that a randomly selected household does not own a dog? |

**Example 6** *Computing Probabilities Using Complements*

The data in Table 4 represent the travel time to work for residents of Hartford County, Connecticut.

**Table 4**

| Travel Time | Frequency |
|---|---|
| Less than 5 minutes | 24,358 |
| 5 to 9 minutes | 39,112 |
| 10 to 14 minutes | 62,124 |
| 15 to 19 minutes | 72,854 |
| 20 to 24 minutes | 74,386 |
| 25 to 29 minutes | 30,099 |
| 30 to 34 minutes | 45,043 |
| 35 to 39 minutes | 11,169 |
| 40 to 44 minutes | 8045 |
| 45 to 59 minutes | 15,650 |
| 60 to 89 minutes | 5451 |
| 90 or more minutes | 4895 |

Data from United States Census Bureau

A) What is the probability that a randomly selected resident has a travel time of 90 or more minutes?

B) What is the probability that a randomly selected resident of Hartford County, Connecticut will have a travel time less than 90 minutes?

## Section 5.3
## Independence and the Multiplication Rule

**Objectives**

❶ Identify Independent Events

❷ Use the Multiplication Rule for Independent Events

❸ Compute At-least Probabilities

---

### *Objective 1: Identify Independent Events*

<u>OBJECTIVE 1, PAGE 1</u>

*Answer the following as you watch the video.*

1) Define independent events and dependent events.

2) Explain why the events "draw a heart" and "roll an even number" are independent.

3) Explain why the events "woman 1 survives the year" and "woman 2 survives the year" are dependent if the two women live in the same complex.

4) When we take a very small sample from a very large finite population, we make the assumption of independence even though the events are technically dependent. State the general rule of thumb for assuming independence.

<u>Objective 1, Page 4</u>
5) Are disjoint events independent?

---

## Objective 2: Use the Multiplication Rule for Independent Events

<u>Objective 2, Page 1</u>
6) State the Multiplication Rule for Independent Events.

<u>Objective 2, Page 2</u>

| Example 1 | Computing Probabilities of Independent Events |
| --- | --- |

In the game of roulette, the wheel has slots numbered 0, 00, and 1 through 36. A metal ball rolls around a wheel until it falls into one of the numbered slots. What is the probability that the ball will land in the slot numbered 17 two times in a row?

<u>Objective 2, Page 3</u>
7) State the Multiplication Rule for $n$ Independent Events.

<u>Objective 2, Page 4</u>

| Example 2 | Life Expectancy |
| --- | --- |

The probability that a randomly selected 24-year-old male will survive the year is 0.9986 according to the National Vital Statistics Report, Vol. 56, No. 9.
A) What is the probability that three randomly selected 24-year-old males will survive the year?

B) What is the probability that twenty randomly selected 24-year-old males will survive the year?

## *Objective 3: Compute At-least Probabilities*

OBJECTIVE 3, PAGE 1

Usually, when computing probabilities involving the phrase *at least*, use the Complement Rule. The phrase *at least* means "greater than or equal to."

OBJECTIVE 3, PAGE 2

---

**Example 3**     *Computing At-least Probabilities*

The probability that a randomly selected female aged 60 years will survive the year is 0.99186 according to the National Vital Statistics Report. What is the probability that at least one of 500 randomly selected 60-year-old females will die during the course of the year?

---

OBJECTIVE 3, PAGE 4

**Summary: Rules of Probability**

**Rule 1** The probability of any event must be between 0 and 1, inclusive. If we let $E$ denote any event, then $0 \le P(E) \le 1$.

**Rule 2** The sum of the probabilities of all outcomes in the sample space must equal 1. That is, if the sample space $S = \{e_1, e_2, \ldots, e_n\}$, then

$$P(e_1) + P(e_2) + \ldots P(e_n) = 1$$

**Rule 3** If $E$ and $F$ are disjoint events, then $P(E \text{ or } F) = P(E) + P(F)$. If $E$ and $F$ are not disjoint events, then $P(E \text{ or } F) = P(E) + P(F) - P(E \text{ and } F)$.

**Rule 4** If $E$ represents any event and $E^C$ represents the complement of $E$, then $P(E^C) = 1 - P(E)$.

**Rule 5** If $E$ and $F$ are independent events, then

$$P(E \text{ and } F) = P(E) \cdot P(F)$$

Notice that *or* probabilities use the Addition Rule, whereas *and* probabilities use the Multiplication Rule. Accordingly, *or* probabilities imply addition, whereas *and* probabilities imply multiplication.

# Section 5.4
# Conditional Probability and the General Multiplication Rule

**Objectives**

❶ Compute Conditional Probabilities

❷ Compute Probabilities Using the General Multiplication Rule

## Objective 1: Compute Conditional Probabilities

OBJECTIVE 1, PAGE 1

1) What does the notation $P(F \mid E)$ represent?

OBJECTIVE 1, PAGE 3

| **Example 1** | ***An Introduction to Conditional Probability*** |

Suppose a single die is rolled. What is the probability that the die comes up three? Now suppose that the die is rolled a second time, but we are told the outcome will be an odd number. What is the probability that the die comes up three?

OBJECTIVE 1, PAGE 5

2) State the Conditional Probability Rule.

OBJECTIVE 1, PAGE 6

**Example 2**     *Conditional Probabilities on Marital Status and Gender*

The data in Table 5 represent the marital status and gender of U.S. residents aged 15 years and older in 2016.

**Table 5**

|  | Males (in millions) | Females (in millions) | Totals (in millions) |
|---|---|---|---|
| Never Married | 44.1 | 39.0 | 83.1 |
| Married | 66.7 | 67.5 | 134.2 |
| Widowed | 3.5 | 11.4 | 14.9 |
| Divorced | 10.7 | 14.8 | 25.5 |
| Totals (in millions) | 125.0 | 132.7 | 257.7 |

A) Compute the probability that a randomly selected individual never married, given that the individual is male.

B) Compute the probability that a randomly selected individual is male, given that the individual never married.

OBJECTIVE 1, PAGE 8

**Example 3**     *Birth Weights of Preterm Babies*

Suppose that 12.2% of all births are preterm. (Preterm means that the gestation period of the pregnancy is less than 37 weeks.) Also, 0.2% of all births result in a preterm baby who weighs 8 pounds, 13 ounces or more. What is the probability that a randomly selected baby weighs 8 pounds, 13 ounces or more, given that the baby is preterm? Is this unusual? Data based on the Vital Statistics Reports.

## Objective 2: *Compute Probabilities Using the General Multiplication Rule*

<u>OBJECTIVE 2, PAGE 1</u>

3) State the General Multiplication Rule.

<u>OBJECTIVE 2, PAGE 2</u>

| **Example 4** | ***Using the General Multiplication Rule*** |

The probability that a driver who is speeding gets pulled over is 0.8. The probability that a driver gets a ticket, given that he or she is pulled over, is 0.9. What is the probability that a randomly selected driver who is speeding gets pulled over and gets a ticket?

<u>OBJECTIVE 2, PAGE 4</u>

| **Example 5** | *Acceptance Sampling* |

Suppose that of 100 circuits sent to a manufacturing plant, 5 are defective. The plant manager receiving the circuits randomly selects two and tests them. If both circuits work, she will accept the shipment. Otherwise, the shipment is rejected. What is the probability that the plant manager discovers at least one defective circuit and rejects the shipment?

**Example 6**    *Favorite Other*

In a study to determine whether preferences for self are more or less prevalent than preferences for others, researchers first asked individuals to identify the person who is most valuable and likeable to you, or favorite other.
Of the 1519 individuals surveyed, 42 had chosen themselves as their favorite other.
Source: Gebauer JE, et al. Self-Love or Other-Love? Explicit Other-Preference but Implicit Self-Preference. PLoS ONE 7(7):e41789. doi:10.1371/journal.prone.0041789

A) Suppose we randomly select 1 of the 1519 individuals surveyed. What is the probability that he or she chose themselves as their favorite other?

B) If two individuals from this group are randomly selected, what is the probability that both chose themselves as their favorite other?

C) Compute the probability of randomly selecting two individuals from this group who selected themselves as their favorite other assuming independence.

If small random samples are taken from large populations without replacement, it is reasonable to assume independence of the events. As a rule of thumb, if the sample size, $n$, is less than 5% of the population size, $N$, we treat the events as independent. That is, if $n < 0.05N$, treat the events as independent.

4) State the definition for independence using conditional probabilities.

# Section 5.5
# Counting Techniques

**Objectives**

❶ Solve Counting Problems Using the Multiplication Rule

❷ Solve Counting Problems Using Permutations

❸ Solve Counting Problems Using Combinations

❹ Solve Counting Problems Involving Permutations with Nondistinct Items

❺ Compute Probabilities Involving Permutations and Combinations

---

## *Objective 1: Solve Counting Problems Using the Multiplication Rule*

<u>OBJECTIVE 1, PAGE 2</u>

| **Example 1** | ***Counting the Number of Possible Meals*** |
|---|---|

The fixed-price dinner at Mabenka Restaurant provides the following choices:
    Appetizer: soup or salad
    Entrée: baked chicken, broiled beef patty, baby beef liver, or roast beef au jus
    Dessert: ice cream or cheesecake
How many different meals can be ordered?

<u>OBJECTIVE 1, PAGE 3</u>
1) State the Multiplication Rule of Counting.

OBJECTIVE 1, PAGE 4

### Example 2   *Counting Airport Codes (Repetition Allowed)*

The International Airline Transportation Association (IATA) assigns three-letter codes to represent airport locations. For example, the code for Fort Lauderdale International Airport is FLL. How many different airport codes are possible?

OBJECTIVE 1, PAGE 5

### Example 3   *Counting (Without Repetition)*

Three members from a 14-member committee are to be randomly selected to serve as chair, vice-chair, and secretary. The first person selected is the chair, the second is the vice-chair, and the third is the secretary. How many different committee structures are possible?

OBJECTIVE 1, PAGE 7

2) Give the definition of *n* factorial.

OBJECTIVE 1, PAGE 10

### Example 4   *The Traveling Salesperson*

You have just been hired as a book representative for Pearson Education.
On your first day, you must travel to seven schools to introduce yourself.
How many different routes are possible?

## Objective 2: Solve Counting Problems Using Permutations

<u>OBJECTIVE 2, PAGE 1</u>
3) State the definition of a permutation.

<u>OBJECTIVE 2, PAGE 2</u>
4) State the formula for the number of permutations of *n* distinct objects taken *r* at a time.

<u>OBJECTIVE 2, PAGE 3</u>

**Example 5**    *Computing Permutations*

Evaluate:
A) $_7P_5$              B) $_5P_5$

<u>OBJECTIVE 2, PAGE 5</u>

**Example 6**    *Betting the Trifecta*

In how many ways can horses in a ten-horse race finish first, second, and third?

## *Objective 3*: *Solve Counting Problems Using Combinations*

<u>OBJECTIVE 3, PAGE 1</u>

5) State the definition of a combination.

<u>OBJECTIVE 3, PAGE 2</u>

| **Example 7** | *Listing* **Combinations** |
|---|---|

Roger, Ken, Tom, and Jay are going to play golf. They will randomly select teams of two players each. List all possible team combinations. That is, list all the combinations of the four people Roger, Ken, Tom, and Jay taken two at a time. What is $_4C_2$?

<u>OBJECTIVE 3, PAGE 4</u>

6) State the formula for the number of combinations of $n$ distinct objects taken $r$ at a time.

<u>OBJECTIVE 3, PAGE 5</u>

| **Example 8** | *Computing Combinations* |
|---|---|

Evaluate:

A) $_4C_1$ 　　　　　　　　B) $_6C_4$ 　　　　　　　　C) $_6C_2$

OBJECTIVE 3, PAGE 7

---

**Example 9**     *Simple Random Samples*

How many different simple random samples of size 4 can be obtained from a population whose size is 20?

---

*Objective 4: Solve Counting Problems Involving Permutations with Nondistinct Items*

OBJECTIVE 4, PAGE 1

---

**Example 10**     *DNA Sequence*

A DNA sequence consists of a series of letters representing a DNA strand that spells out the genetic code. There are four possible letters (A, C, G, and T), each representing a specific nucleotide base in the DNA strand (adenine, cytosine, guanine, and thymine, respectively).
How many distinguishable sequences can be formed using two As, two Cs, three Gs, and one T?

---

OBJECTIVE 4, PAGE 2

7) State the formula for permutations with nondistinct items.

Copyright © 2019 Pearson Education, Inc.

**Example 11**     *Arranging Flags*

How many different vertical arrangements are there of ten flags if five are white, three are blue, and two are red?

**Summary: Combinations and Permutations**

|  | Description | Formula |
|---|---|---|
| **Combination** | The selection of $r$ objects from a set of $n$ different objects when the order in which the objects are selected does not matter (so $AB$ is the same as $BA$) and an object cannot be selected more than once (repetition is not allowed) | $_nC_r = \dfrac{n!}{(n-r)!}$ |
| **Permutation of Distinct Items with Replacement** | The selection of $r$ objects from a set of $n$ different objects when the order in which the objects are selected matters (so $AB$ is different from $BA$) and an object may be selected more than once (repetition is allowed) | $n^r$ |
| **Permutation of Distinct Items without Replacement** | The selection of $r$ objects from a set of $n$ different objects when the order in which the objects are selected matters so ($AB$ is different from $BA$) and an object cannot be selected more than once (repetition is not allowed) | $_nP_r = \dfrac{n!}{r!(n-r)!}$ |
| **Permutation of Nondistinct Items without Replacement** | The number of ways $n$ objects can be arranged (order matters) in which there are $n_1$ of one kind, $n_2$ of a second kind, ..., and $n_k$ of a $k$th kind, where $n = n_1 + n_2 + ... + n_k$ | $\dfrac{n!}{n_1! \cdot n_2! \cdot ... \cdot n_k!}$ |

*Objective 5: Compute Probabilities Involving Permutations and Combinations*

**Example 12**   *Winning the Lottery*

In the Illinois Lottery, an urn contains balls numbered 1 to 52. From this urn, six balls are randomly chosen without replacement. For a $1 bet, a player chooses two sets of six numbers. To win, all six numbers must match those chosen from the urn. The order in which the balls are picked does not matter. What is the probability of winning the lottery?

**Example 13**   *Acceptance Sampling*

A shipment of 120 fasteners that contains 4 defective fasteners was sent to a manufacturing plant. The plant's quality control manager randomly selects and inspects 5 fasteners. What is the probability that exactly 1 of the inspected fasteners is defective?

## Section 5.6
## Simulation

**Objective**

❶ Use Simulation to Obtain Probabilities

---

### Objective 1: Use Simulation to Obtain Probabilities

<u>OBJECTIVE 1, PAGE 1</u>

*Answer the following while watching the video.*

1) List two historical uses of simulation.

<u>OBJECTIVE 1, PAGE 2</u>

| **Example 1** | *Getting Out of Jail in Monopoly* |

In the board game Monopoly, a player can get out of jail in one of three ways.

   1. The player pays a $50 fine.
   2. The player uses a "Get Out of Jail" card.
   3. The player rolls doubles.

If the player does not roll doubles after three rolls, the player must pay the $50 fine. Use simulation to determine the probability that a player will not roll doubles after three consecutive rolls.

<u>OBJECTIVE 1, PAGE 5</u>

When collecting data for an observational study, it is important that individuals are randomly selected to be in the study. This allows the results of the study to be extended to the population from which the individuals were randomly selected.

When collecting data for a designed experiment, it is important that the individuals are randomly assigned to the various treatment groups in the study. This allows us to make statements of causation between the levels of treatment and the response variable in the study.

<u>OBJECTIVE 1, PAGE 6</u>

**Example 2**     *Random Selection–Qualitative Response*

Unplugging refers to eliminating the use of social media, cell phones, and other technology. According to Harris Interactive, the proportion of adult Americans (aged 18 years or older) who attempt to "unplug" at least once a week is 0.45. There are approximately 241,000,000 Americans aged 18 years or older in the United States.

A) Simulate obtaining a simple random sample of size 500 from the population. How many of the individuals sampled unplug? How many do not unplug? What proportion unplug at least once a week?

B) Simulate obtaining a second simple random sample of size 500 from the population. How many of the individuals sampled unplug? How many do not unplug? What proportion unplug at least once a week? Why will the results of the first sample differ from those in the second sample?

C) Now simulate obtaining at least 2000 more simple random samples of size 500 from the population. Based on the simulation, what is the probability of obtaining a random sample where the proportion who unplug at least once a week is greater than 0.50? Would it be unusual to obtain a sample proportion greater than 0.5 from this population? Explain.

## Section 5.7
## Putting It Together: Which Method Do I Use?

**Objectives**

❶ Determine the Appropriate Probability Rule to Use

❷ Determine the Appropriate Counting Technique to Use

___

### *Objective 1: Determine the Appropriate Probability Rule to Use*

OBJECTIVE 1, PAGE 1
**Flowchart for Probability Rules**

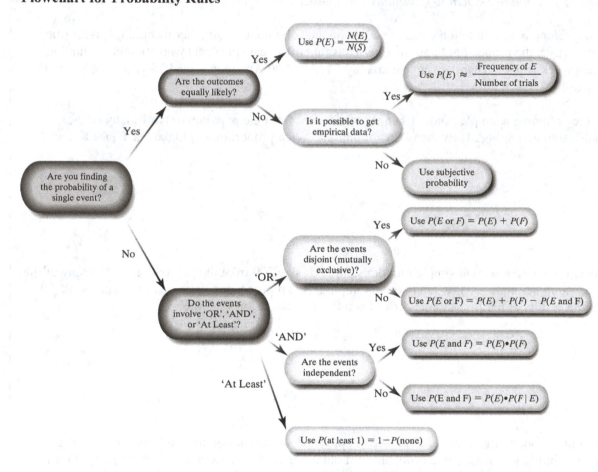

1) What are three options when finding the probability of a single event?

OBJECTIVE 1, PAGE 1 (CONTINUED)

2) What must you determine when working with events involving the word "AND"?

3) What must you determine when working with events involving the word "OR"?

OBJECTIVE 1, PAGE 2

---

**Example 1**     *Probability: Which Rule Do I Use?*

In the game show Deal or No Deal?, a contestant is presented with 26 suitcases that contain amounts ranging from $0.01 to $1,000,000. The contestant must pick an initial case that is set aside as the game progresses. The amounts are randomly distributed among the suitcases prior to the game as shown in Table 7. What is the probability that the contestant picks a case worth at least $100,000?

**Table 7**

| Prize | Number of Suitcases |
|---|---|
| $0.01–$100 | 8 |
| $200–$1000 | 6 |
| $5000–$50,000 | 5 |
| $100,000–$1,000,000 | 7 |

---

OBJECTIVE 1, PAGE 3

---

**Example 2**     *Probability: Which Rule Do I Use?*

According to a Harris poll, 14% of adult Americans have one or more tattoos, 50% have pierced ears, and 65% of those with one or more tattoos also have pierced ears. What is the probability that a randomly selected adult American has one or more tattoos and pierced ears?

---

## Objective 2: Determine the Appropriate Counting Technique to Use

**Flowchart for Counting Techniques**

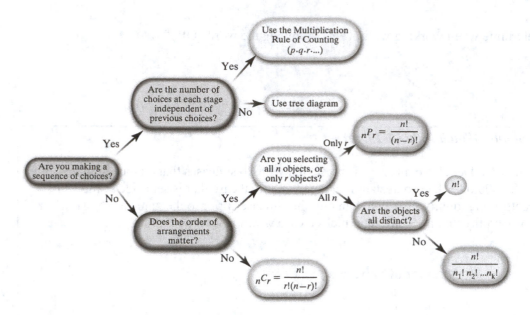

4) What counting techniques can be used when working with a sequence of choices? Explain when to use each strategy.

5) What counting techniques can be used when working with the number of arrangements of items? Explain when to use each strategy.

OBJECTIVE 2, PAGE 2

---

**Example 3**     *Counting: Which Technique Do I Use?*

The Hazelwood city council consists of 5 men and 4 women. How many different subcommittees can be formed that consist of 3 men and 2 women?

---

OBJECTIVE 2, PAGE 3

---

**Example 4**     *Counting: Which Technique Do I Use?*

The Daytona 500, the season opening NASCAR event, has 43 drivers in the race. In how many different ways could the top four finishers (first, second, third, and fourth place) occur?

---

# Chapter 6 – Discrete Probability Distributions

## OUTLINE

**Putting It Together**

Recall, the probability of an event is the long-term proportion with which the event is observed. That is, if we conduct an experiment 1000 times and observe an outcome 300 times, we estimate that the probability of the outcome is 300/1000 = 0.3. The more times we conduct the experiment, the more accurate this empirical probability will be. This is the Law of Large Numbers. We can also use counting techniques to obtain theoretical probabilities if the outcomes in the experiment are equally likely. This is called classical probability.

A probability model lists the possible outcomes of a probability experiment and each outcome's probability. A probability model must satisfy the rules of probability. In particular, all probabilities must be between 0 and 1, inclusive, and the sum of the probabilities must equal 1.

Now we introduce probability models for random variables. A random variable is a numerical measure of the outcome to a probability experiment. So, rather than listing specific outcomes of a probability experiment, such as heads or tails, we might list the number of heads obtained in three flips of a coin. We begin by discussing random variables and describe the distribution of discrete random variables (shape, center, and spread). We then discuss two specific discrete probability distributions: the binomial probability distribution and the Poisson probability distribution.

## Section 6.1
## Discrete Random Variables

**Objectives**

❶ Distinguish between Discrete and Continuous Random Variables

❷ Identify Discrete Probability Distributions

❸ Graph Discrete Probability Distributions

❹ Compute and Interpret the Mean of a Discrete Random Variable

❺ Interpret the Mean of a Discrete Random Variable as an Expected Value

❻ Compute the Standard Deviation of a Discrete Random Variable

---

### Objective 1: Distinguish between Discrete and Continuous Random Variables

OBJECTIVE 1, PAGE 1

1) Give the definition of a random variable.

OBJECTIVE 1, PAGE 2

2) There are two types of random variables, discrete and continuous. Explain the difference between them.

OBJECTIVE 1, PAGE 4

| **Example 1** | ***Distinguishing between Discrete and Continuous Random Variables*** |

Determine whether the random variable is a discrete random variable or a continuous random variable.
A) The number of As earned in a section of statistics with 15 students enrolled
B) The number of cars that travel through a McDonald's drive-through in the next hour
C) The speed of the next car that passes a state trooper

## *Objective 2: Identify Discrete Probability Distributions*

<u>OBJECTIVE 2, PAGE 1</u>
3) Give the definition of a probability distribution.

<u>OBJECTIVE 2, PAGE 2</u>

**Example 2** *A Discrete Probability Distribution*

Suppose we ask a basketball player to shoot three free throws. Let the random variable $X$ represent the number of shots made; so $x = 0$, 1, 2, or 3. Table 1 shows a probability distribution for the random variable $X$.

**Table 1**

| $x$ | $P(x)$ |
|-----|--------|
| 0 | 0.01 |
| 1 | 0.10 |
| 2 | 0.38 |
| 3 | 0.51 |

A) What does the notation $P(x)$ represent?

B) Explain what $P(3) = 0.51$ represents.

<u>OBJECTIVE 2, PAGE 3</u>
4) State the rules for a discrete probability distribution.

Chapter 6: Discrete Probability Distributions

| Example 3 | *Identifying Discrete Probability Distributions* |

Which of the following is a discrete probability distribution?

A)

| x | P(x) |
|---|------|
| 0 | 0.16 |
| 1 | 0.18 |
| 2 | 0.22 |
| 3 | 0.10 |
| 4 | 0.30 |
| 5 | 0.01 |

B)

| x | P(x) |
|---|------|
| 0 | 0.16 |
| 1 | 0.18 |
| 2 | 0.22 |
| 3 | 0.10 |
| 4 | 0.30 |
| 5 | −0.01 |

C)

| x | P(x) |
|---|------|
| 0 | 0.16 |
| 1 | 0.18 |
| 2 | 0.22 |
| 3 | 0.10 |
| 4 | 0.30 |
| 5 | 0.04 |

## Objective 3: Graph Discrete Probability Distributions

5) In the graph of a discrete probability distribution, what do the horizontal axis and the vertical axis represent?

6) When graphing a discrete probability distribution, how do we emphasize that the data is discrete?

OBJECTIVE 3, PAGE 2

---

**Example 4**     *Graph a Discrete Probability Distribution*

Graph the discrete probability distribution given in Table 1.

**Table 1**

| $x$ | $P(x)$ |
|-----|--------|
| 0 | 0.01 |
| 1 | 0.10 |
| 2 | 0.38 |
| 3 | 0.51 |

---

OBJECTIVE 3, PAGE 3

Graphs of discrete probability distributions help determine the shape of the distribution.
Recall that we describe distributions as skewed left, skewed right, or symmetric.

---

*Objective 4: Compute and Interpret the Mean of a Discrete Random Variable*

OBJECTIVE 4, PAGE 1

*Watch the video to learn about the derivation of the formula for finding the mean of a discrete random variable.*

OBJECTIVE 4, PAGE 2

7) State the formula for the mean of a discrete random variable.

OBJECTIVE 4, PAGE 3

| Example 5 | Computing the Mean of a Discrete Random Variable |
|---|---|

Compute the mean of the discrete probability distribution given in Table 1.

**Table 1**

| x | P(x) |
|---|------|
| 0 | 0.01 |
| 1 | 0.10 |
| 2 | 0.38 |
| 3 | 0.51 |

OBJECTIVE 4, PAGE 4

*Answer the following after watching the video.*

8) As the number of repetitions of the experiments increases, what does the mean value of the *n* trials approach?

9) As the number of repetitions of the experiments increases, what happens to the difference between the mean outcome and the mean of the probability distribution?

OBJECTIVE 4, PAGE 5

| Example 6 | Interpretation of the Mean of a Discrete Random Variable |
|---|---|

The basketball player from Example 2 is asked to shoot three free throws 100 times. Compute the mean number of free throws made.

In each simulation, what value is the graph (that shows the mean number of free throws made) drawn towards?

*Objective 5: Interpret the Mean of a Discrete Random Variable as an Expected Value*

<u>OBJECTIVE 5, PAGE 1</u>

Because the mean of a random variable represents what we would expect to happen in the long run, it is also called the expected value, $E(X)$. The interpretation of the expected value is the same as the interpretation of the mean of a discrete random variable.

<u>OBJECTIVE 5, PAGE 2</u>

---

**Example 7**    *Computing the Expected Value of a Discrete Random Variable*

A term life insurance policy will pay a beneficiary a certain sum of money upon the death of the policy holder. These policies have premiums that must be paid annually. Suppose a life insurance company sells a $250,000 one-year term life insurance policy to a 49-year-old female for $530. According to the National Vital Statistics Report, Vol. 47, No. 28, the probability that the female will survive the year is 0.99791. Compute the expected value of this policy to the insurance company.

---

*Objective 6: Compute the Standard Deviation of a Discrete Random Variable*

<u>OBJECTIVE 6, PAGE 1</u>

10) State the formula for computing the standard deviation of a discrete random variable.

**Example 8**    *Computing the Standard Deviation of a Discrete Random Variable*

Compute the standard deviation of the discrete random variable given in Table 1.

**Table 1**

| x | P(x) |
|---|---|
| 0 | 0.01 |
| 1 | 0.10 |
| 2 | 0.38 |
| 3 | 0.51 |

The variance of the discrete random variable, $\sigma_X^2$, is the value under the square root in the computation of the standard deviation.

## Section 6.2
## The Binomial Probability Distribution

**Objectives**

❶ Determine Whether a Probability Experiment is a Binomial Experiment

❷ Compute Probabilities of Binomial Experiments

❸ Compute the Mean and Standard Deviation of a Binomial Random Variable

❹ Graph a Binomial Probability Distribution

### Objective 1: Determine Whether a Probability Experiment is a Binomial Experiment

OBJECTIVE 1, PAGE 1

The binomial probability distribution is a discrete probability distribution that describes probabilities for experiments in which there are two mutually exclusive (disjoint) outcomes. These two outcomes are generally referred to as success (such as making a free throw) and failure (such as missing a free throw). Experiments in which only two outcomes are possible are referred to as binomial experiments, provided that certain criteria are met.

OBJECTIVE 1, PAGE 2

*Answer the following as you watch the video.*

1) What are the four criteria for a binomial experiment?

2) What do $n$, $p$, and $1 - p$ represent when working with a binomial probability distribution?

OBJECTIVE 1, PAGE 2 (CONTINUED)
3) If $X$ is a binomial random variable that denotes the number of successes in $n$ independent trials of an experiment, what are the possible values of $X$?

OBJECTIVE 1, PAGE 3

| **Example 1** | *Identifying Binomial Experiments* |
|---|---|

Determine which of the following probability experiments qualify as binomial experiments. For those that are binomial experiments, identify the number of trials, probability of success, probability of failure, and possible values of the random variable X.

A) An experiment in which a basketball player who historically makes 80% of his free throws is asked to shoot three free throws and the number of free throws made is recorded.

B) According to a recent Harris Poll, 28% of Americans state that chocolate is their favorite flavor of ice cream. Suppose a simple random sample of size 10 is obtained and the number of Americans who choose chocolate as their favorite ice cream flavor is recorded.

C) A probability experiment in which three cards are drawn from a deck without replacement and the number of aces is recorded.

## Objective 2: Compute Probabilities of Binomial Experiments

OBJECTIVE 2, PAGE 1

*Watch the video to learn how the binomial probability distribution function is developed.*

OBJECTIVE 2, PAGE 2

4) In the formula, what does $_4C_1$ represent?

5) In the formula, what do 0.07 and 1 represent?

6) In the formula, what do 0.93 and 3 represent?

OBJECTIVE 2, PAGE 3

7) State the Binomial Probability Distribution Function (pdf).

OBJECTIVE 2, PAGE 4

8) Fill in the math symbol that is associated with the given phrase.

| Phrase | Math Symbol |
| --- | --- |
| at least or no less than or greater than or equal to | |
| more than or greater than | |
| fewer than or less than | |
| no more than or at most or less than or equal to | |
| exactly or equals or is | |

OBJECTIVE 2, PAGE 5

---

**Example 2**    *Using the Binomial Probability Distribution Function*

According to CTIA, 55% of all U.S. households are wireless-only households (no landline).

A) What is the probability of obtaining exactly ten wireless-only households based on a random sample of fifteen households?

B) What is the probability of obtaining fewer than three wireless-only households based on a random sample of fifteen households?

C) What is the probability of obtaining at least three wireless-only households based on a random sample of fifteen households?

D) What is the probability of obtaining between five and seven, inclusive, wireless-only households based on a random sample of twenty households?

---

## Objective 3: Compute the Mean and Standard Deviation of a Binomial Random Variable

OBJECTIVE 3, PAGE 1
9) State the formulas for the mean (or expected value) and standard deviation of a binomial random variable.

| Example 3 | *Finding the Mean and Standard Deviation of a Binomial Random Variable* |

According to CTIA, 55% of all U.S. households are wireless-only households. In a simple random sample of 500 households, determine the mean and standard deviation number of wireless-only households.

## Objective 4: Graph a Binomial Probability Distribution

To graph a binomial probability distribution, first find the probabilities for each possible value of the random variable. Then follow the same approach as was used to graph discrete probability distributions.

| Example 4 | *Graph a Binomial Probability Distribution* |

A) Graph a binomial probability distribution with n = 10 and p = 0.2. Comment on the shape of the distribution.
B) Graph a binomial probability distribution with n = 10 and p = 0.5. Comment on the shape of the distribution.
C) Graph a binomial probability distribution with n = 10 and p = 0.8. Comment on the shape of the distribution.

OBJECTIVE 4, PAGE 3

10) What is the shape of the binomial probability distribution if $p < 0.5$, if $p = 0.5$, and if $p > 0.5$?

OBJECTIVE 4, PAGE 5

📱 *Answer the following after Activity 1: The Role of n, the Number of Trials of a Binomial Experiment, on Distribution Shape*

11) As *n* increases, describe what happens to the shape of a binomial probability distribution.

OBJECTIVE 4, PAGE 6

12) Under what conditions will a binomial probability distribution be approximately bell-shaped?

13) Explain how to determine if an observation in a binomial experiment is unusual.

OBJECTIVE 4, PAGE 7

| **Example 5** | ***Using the Mean, Standard Deviation, and Empirical Rule to Check for Unusual Results in a Binomial Experiment*** |
|---|---|

According to CTIA, 55% of all U.S. households are wireless-only households. In a simple random sample of 500 households, 301 were wireless-only. Is this result unusual?

## Section 6.3
## The Poisson Probability Distribution

**Objectives**

&#10112; Determine Whether a Probability Experiment Follows a Poisson Process

&#10113; Compute Probabilities of a Poisson Random Variable

&#10114; Find the Mean and Standard Deviation of a Poisson Random Variable

---

### Objective 1: Determine Whether a Probability Experiment Follows a Poisson Process

<u>OBJECTIVE 1, PAGE 1</u>

1) For what situations in the Poisson probability distribution used?

<u>OBJECTIVE 1, PAGE 2</u>

> **Example 1**   *Illustrating a Poisson Process*
>
> A McDonald's® manager knows from experience that cars arrive at the drive-through at an average rate of two cars per minute between the hours of 12:00 noon and 1:00 PM. The random variable X, the number of cars that arrive between 12:20 and 12:40, follows a Poisson process.

<u>OBJECTIVE 1, PAGE 3</u>

2) Under what conditions does a random variable *X* follow a Poisson process?

## Objective 2: Compute Probabilities of a Poisson Random Variable

<u>OBJECTIVE 2, PAGE 1</u>

3) State the Poisson Probability Distribution Function.

<u>OBJECTIVE 2, PAGE 2</u>

**Example 2**     *Computing Probabilities of a Poisson Process*

A McDonald's manager knows that cars arrive at the drive-through at the average rate of two cars per minute between the hours of 12 noon and 1:00 PM. Find the following probabilities.

A) Find the probability that exactly six cars arrive between 12 noon and 12:05 PM.

B) Find the probability that fewer than six cars arrive between 12 noon and 12:05 PM.

C) Find the probability that at least six cars arrive between 12 noon and 12:05 PM.

## *Objective 3: Find the Mean and Standard Deviation of a Poisson Random Variable*

OBJECTIVE 3, PAGE 1

4) State the formula for the mean and standard deviation of a Poisson random variable.

OBJECTIVE 3, PAGE 2

5) Restate the Poisson probability distribution function in terms of its mean.

OBJECTIVE 3, PAGE 3

| **Example 3** | **Beetles and the Poisson Distribution** |

A biologist performs an experiment in which 2000 Asian beetles are allowed to roam in an enclosed area of 1000 square feet. The area is divided into 200 subsections of 5 square feet each.

A) If the beetles spread themselves evenly throughout the enclosed area, how many beetles would you expect in each subsection?

B) What is the standard deviation of X, the number of beetles in a particular subsection?

C) What is the probability of finding exactly eight beetles in a particular subsection?

D) Would it be unusual to find more than 16 beetles in a particular subsection?

# Chapter 7 – The Normal Probability Distribution

## OUTLINE

**Putting It Together**

In Chapter 6, we introduced discrete probability distributions. We computed probabilities using probability distribution functions. However, we could also determine the probability of any discrete random variable from its probability histogram. For example, the figure below shows the probability histogram for the binomial random variable $X$ with $n = 5$ and $p = 0.35$.

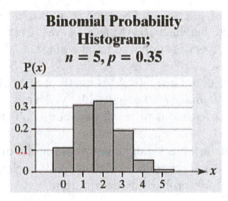

From this probability histogram, we see that $P(1) \approx 0.31$. Notice that the width of each rectangle in the probability histogram is 1. Since the area of a rectangle equals height times width, we can think of $P(1)$ as the area of the rectangle corresponding to $x = 1$. Thinking of probability in this way makes the transition from computing discrete probabilities to finding continuous probabilities much easier.

In this chapter, we discuss two continuous distributions: the *uniform distribution* and the *normal distribution*. Most of the discussion will focus on the normal distribution, which has many applications.

## Section 7.1
## Properties of the Normal Distribution

**Objectives**

&#10112; Use the Uniform Probability Distribution

&#10113; Graph a Normal Curve

&#10114; State the Properties of the Normal Curve

&#10115; Explain the Role of Area in the Normal Density Function

---

### Objective 1: Use the Uniform Probability Distribution

OBJECTIVE 1, PAGE 1

We discuss a uniform distribution to see the relation between area and probability.

OBJECTIVE 1, PAGE 2

| **Example 1**     *The Uniform Distribution* |
| --- |
| Assume that United Parcel Service is supposed to deliver a package to your front door and the arrival time is somewhere between 10 AM and 11 AM. Let the random variable $X$ represent the time from 10 AM when the delivery is supposed to take place. |
| The delivery could be at 10 AM ($x = 0$) or at 11 AM ($x = 60$), with all one-minute intervals of time between $x = 0$ and $x = 60$ equally likely. That is to say, your package is just as likely to arrive between 10:15 and 10:16 as it is to arrive between 10:40 and 10:41. |
| The random variable $X$ can be any value in the interval from 0 to 60, that is, $0 \leq X \leq 60$. Because any two intervals of equal length between 0 and 60, inclusive, are equally likely, the random variable $X$ is said to follow a uniform probability distribution. |

OBJECTIVE 1, PAGE 3

*Answer the following after watching the video.*

1) What two properties must a probability density function (pdf) satisfy?

OBJECTIVE 1, PAGE 3 (CONTINUED)

2) If the possible values of a uniform density function go from 0 through *n*, what is the height of the rectangle?

3) What does the area under the graph of a probability density function over an interval represent?

---

## Objective 2: Graph a Normal Curve

OBJECTIVE 2, PAGE 1

Not all continuous random variables follow a uniform distribution.

In Figure 1, as the class width of the histogram decreases, the histogram becomes closely approximated by the smooth red curve. For this reason, we can use the curve to *model* the probability distribution of this continuous random variable.

OBJECTIVE 2, PAGE 2

4) What does it mean to say that a continuous random variable is normally distributed?

OBJECTIVE 2, PAGE 3

*Answer the following while watching the video.*

5) What value of *x* is associated with the peak of a normal curve?

6) What values of *x* are associated with the inflection points of a normal curve?

7) Sketch and label the graph from Figure 2.

*Answer the following after Activity 1: The Role of $\mu$ and $\sigma$ in a Normal Curve.*

8) What happens to the graph as the mean increases? What happens to the graph as the mean decreases?

9) What happens to the graph as the standard deviation increases? What happens to the graph as the standard deviation decreases?

## Objective 3: State the Properties of the Normal Curve

<u>OBJECTIVE 3, PAGE 1</u>
10) State the seven properties of the normal curve.

## Objective 4: Explain the Role of Area in the Normal Density Function

<u>OBJECTIVE 4, PAGE 1</u>
*Watch the video to see an example of a normally distributed random variable.*

The area under the normal curve can be used to model the probability histogram and the actual proportion in a given interval.

11) Suppose that a random variable $X$ is normally distributed with mean $\mu$ and standard deviation $\sigma$. Give two representations for the area under the normal curve for any interval of values of the random variable $X$.

---

**Example 2**     *Interpreting the Area Under a Normal Curve*

The serum total cholesterol for males 20 to 29 years old is approximately normally distributed with mean $\mu = 180$ and standard deviation $\sigma = 36.2$, based on data obtained from the National Health and Nutrition Examination Survey.

A) Draw a normal curve with the parameters labeled.
B) An individual with total cholesterol greater than 200 is considered to have high cholesterol. Shade the region under the normal curve to the right of $x = 200$.
C) Suppose that the area under the normal curve to the right of $x = 200$ is 0.2903. (You will learn how to find this area in the next section.) Provide two interpretations of this result.

---

**178**

| Section 7.2 |
| :---: |
| **Applications of the Normal Distribution** |

**Objectives**

❶ Find and Interpret the Area under a Normal Curve

❷ Find the Value of a Normal Random Variable

---

*Objective 1: Find and Interpret the Area under a Normal Curve*

OBJECTIVE 1, PAGE 1

1) Suppose that the random variable $X$ is normally distributed with mean $\mu$ and standard deviation $\sigma$. .

Explain the distribution of the random variable $Z = \dfrac{X - \mu}{\sigma}$. What is the name for the random variable $Z$?

OBJECTIVE 1, PAGE 2

2) Explain how to find the area to the left of $x$ for a normally distributed random variable $X$, using Table V.

OBJECTIVE 1, PAGE 3

Answer the following after watching the video.

3) Explain how to find the area to the right of $x$ for a normally distributed random variable $X$, using Table V.

> **Example 1**     *Finding and Interpreting Area Under a Normal Curve*
>
> A pediatrician obtains the heights of her 200 three-year-old female patients. The heights are approximately normally distributed, with mean 38.72 inches and standard deviation 3.17 inches. Use the normal model to determine the proportion of the 3-year-old females who have a height less than 35 inches.

Note that the proportion of 3-year-old females who are shorter than 35 inches according to the normal model is close to the actual results. The normal curve accurately models the heights.

Because the area under the normal curve represents a proportion, we can also use the area to find percentile ranks of scores.

> **Example 2**     *Finding and Interpreting Area Under a Normal Curve*
>
> A pediatrician obtains the heights of her 200 three-year-old female patients. The heights are approximately normally distributed, with mean 38.72 inches and standard deviation 3.17 inches. Use the normal model to determine the probability that a randomly selected 3-year-old female is between 35 and 40 inches tall, inclusive. That is, find $P(35 \le X \le 40)$.

*Answer the following after watching the video.*

4) Summarize the methods for finding the area to the left of *x*, the area to the right of *x*, and the area between $x_1$ and $x_2$.

**Area to the Left of *x***

**Area to the Right of *x***

**Area Between $x_1$ and $x_2$**

## Objective 2: Find the Value of a Normal Random Variable

OBJECTIVE 2, PAGE 1
Often, we do not want to find the proportion, probability, or percentile given a value of a normal random variable. Rather, we want to find the value of a normal random variable that corresponds to a certain proportion, probability, or percentile. For example, we might want to know the height of a 3-year-old girl who is at the 20th percentile. Or we might want to know the scores on a standardized exam that separate the middle 90% of scores from the bottom and top 5%.

OBJECTIVE 2, PAGE 2

**Example 3**   *Finding the Value of a Normal Random Variable*

The heights of a pediatrician's 3-year-old female patients are approximately normally distributed, with mean 38.72 inches and standard deviation 3.17 inches. Find the height of a 3-year-old female at the 20th percentile.

OBJECTIVE 2, PAGE 4

**Example 4**   *Finding the Value of a Normal Random Variable*

The scores earned on the mathematics portion of the SAT, a college entrance exam, are approximately normally distributed with mean 516 and standard deviation 116. What scores separate the middle 90% of test takers from the bottom and top 5%? In other words, find the 5th and 95th percentiles.
Data from The College Board

<u>OBJECTIVE 2, PAGE 6</u>

5) What does the notation $z_\alpha$ represent?

<u>OBJECTIVE 2, PAGE 7</u>

| **Example 5**     *Finding the Value of* $z_\alpha$ |
|---|
| Find the value of $z_{0.10}$  |

<u>OBJECTIVE 2, PAGE 9</u>

6) For any continuous random variable, what is the probability of observing a specific value of the random variable?

Since the probability of observing a specific value of a continuous random variable is 0, the following probabilities are equivalent.

$$P(a < X < b) = P(a \le X < b) = P(a < X \le b) = P(a \le X \le b)$$

## Section 7.3
## Assessing Normality

**Objective**

❶ Use Normal Probability Plots to Assess Normality

*Objective 1: Use Normal Probability Plots to Assess Normality*

OBJECTIVE 1, PAGE 1

1) What is a normal score?

2) What is a normal probability plot?

3) List the four steps for drawing a normal probability plot by hand.

OBJECTIVE 1, PAGE 4

The idea behind finding the expected z-score is, if the data come from a normally distributed population, we could predict the area to the left of each data value.

OBJECTIVE 1, PAGE 5

4) If sample data are taken from a population that is normally distributed, how will the normal probability plot appear?

5) Explain how to determine if a normal probability plot is "linear enough".

OBJECTIVE 1, PAGE 6

**Example 1**    ***Drawing a Normal Probability Plot by Hand***

The data in Table 2 represent the finishing time (in seconds) for six randomly selected races of a greyhound named Barbies Bomber in the 5/16-mile race at Greyhound Park in Dubuque, Iowa. Is there evidence to support the belief that the variable "finishing time" is normally distributed?

**Table 2**

| | |
|------|------|
| 31.35 | 32.52 |
| 32.06 | 31.26 |
| 31.91 | 32.37 |

Data from Greyhound Park, Dubuque, IA

OBJECTIVE 1, PAGE 8
Typically, normal probability plots are drawn using either a graphing calculator with advanced statistical features or statistical software such as StatCrunch.

OBJECTIVE 1, PAGE 9

**Example 2**     *Drawing a Normal Probability Plot Using Technology*

Draw a normal probability plot of the Barbies Bomber data in Table 2 using technology. Is there evidence to support the belief that the variable "finishing time" is normally distributed?

**Table 2**

| | |
|---|---|
| 31.35 | 32.52 |
| 32.06 | 31.26 |
| 31.91 | 32.37 |

Data from Greyhound Park, Dubuque, IA

**Example 3**    *Assessing Normality*

The data in Table 4 represent the time 100 randomly selected riders spent waiting in line (in minutes) for the Demon Roller Coaster. Is the random variable "wait time" normally distributed?

**Table 4**

| | | | | | | | | | | | | | | | | |
|---|---|---|---|---|---|---|---|---|---|---|---|---|---|---|---|---|
| 7 | 3 | 5 | 107 | 8 | 37 | 16 | 41 | 7 | 25 | 22 | 19 | 1 | 40 | 1 | 29 | 93 |
| 33 | 76 | 14 | 8 | 9 | 45 | 15 | 81 | 94 | 10 | 115 | 18 | 0 | 18 | 11 | 60 | 34 |
| 30 | 6 | 21 | 0 | 86 | 6 | 11 | 1 | 1 | 3 | 9 | 79 | 41 | 2 | 9 | 6 | 19 |
| 4 | 3 | 2 | 7 | 18 | 0 | 93 | 68 | 6 | 94 | 16 | 13 | 24 | 6 | 12 | 121 | 30 |
| 35 | 39 | 9 | 15 | 53 | 9 | 47 | 5 | 55 | 64 | 51 | 80 | 26 | 24 | 12 | 0 | |
| 94 | 18 | 4 | 61 | 38 | 38 | 21 | 61 | 9 | 80 | 18 | 21 | 8 | 14 | 47 | 56 | |

## Section 7.4
## The Normal Approximation to the Binomial Probability Distribution

**Objective**

❶ Approximate Binomial Probabilities Using the Normal Distribution

---

### *Objective 1: Approximate Binomial Probabilities Using the Normal Distribution*

INTRODUCTION, PAGE 1

*Answer the following after watching the video.*

1) What are the three criteria for a binomial probability experiment?

2) Under what conditions will a binomial random variable be approximately normally distributed?

OBJECTIVE 1, PAGE 1

3) If the binomial random variable $X$ is approximately distributed, state the formulas for its mean and standard deviation.

OBJECTIVE 1, PAGE 2

4) If $n = 40$ and $p = 0.5$, we can use a normal model because $np(1-p) = 40(0.5)(0.5) = 10$.

Compute $\mu_X$ and $\sigma_X$.

OBJECTIVE 1, PAGE 4

To approximate the probability of a specific value of the binomial random variable, such as P(18), find the area under the normal curve from x=17.5 to x=18.5. We add and subtract 0.5 from x=18 as a correction for continuity because we are using a continuous density function to approximate a discrete probability.

OBJECTIVE 1, PAGE 5

Watch the video that summarizes the various corrections for continuity.

OBJECTIVE 1, PAGE 6

5) What is the continuity correction in each of the following cases?

A) $P(X = a)$

B) $P(X \leq a)$

C) $P(X \geq a)$

D) $P(a \leq X \leq b)$

OBJECTIVE 1, PAGE 9

**Example 1** *The Normal Approximation to a Binomial Random Variable*

According to the American Red Cross, 7% of people in the United States have blood type O-negative. What is the probability that in a simple random sample of 500 people in the United States fewer than 30 have blood type O-negative?

OBJECTIVE 1, PAGE 10

Note that the approximate result using the normal model is only off by 0.0007 from the exact probability computed using technology. Also, notice the shape of the distribution in the StatCrunch output.

**Example 2** *A Normal Approximation to the Binomial*

According to the Gallup Organization, 65% of adult Americans are in favor of the death penalty for individuals convicted of murder. Erica selects a random sample of 1000 adult Americans in Will County, Illinois, and finds that 630 of them are in favor of the death penalty for individuals convicted of murder.

A) According to the Gallup Organization, 65% of adult Americans are in favor of the death penalty for individuals convicted of murder. Erica selects a random sample of 1000 adult Americans in Will County, Illinois, and finds that 630 of them are in favor of the death penalty for individuals convicted of murder.

B) Does the result from part (A) contradict the Gallup Organization's findings? Explain.

# Chapter 8 – Sampling Distributions

**OUTLINE**

**Putting It Together**

In chapters 6 and 7, we learned about random variables and their probability distributions. In this chapter, we continue our discussion of probability distributions where statistics, such as $\overline{x}$, will be the random variable. Statistics are random variables because the value of a statistic varies from sample to sample. For this reason, statistics have probability distributions associated with them. For example, there is a probability distribution for the sample mean, sample proportion, and so on. We use probability distributions to make probability statements regarding the statistic. In this chapter, we examine the shape, center, and spread of statistics such as $\overline{x}$.

## Section 8.1
## Distribution of the Sample Mean

**Objectives**

❶ Describe the Distribution of the Sample Mean: Normal Population

❷ Describe the Distribution of the Sample Mean: Non-normal Population

INTRODUCTION, PAGE 1

*Watch the video for an overview of where we have been and where we are going in the course.*
In Chapters 1 through 4 we learned how to identify the research objective (Step 1 of the statistical process) as well as collect (Step 2) and describe data (Step 3). In Chapters 5 through 7 we developed the skills that allow us to perform inference (Step 4). Because it is difficult to gain access to populations, the data found in Step 2 is often from a sample. Sample data are used to make inferences about the population. For example, we might compute the mean of a sample and use this information to draw conclusions regarding the population mean. The rest of this course focuses on how sample data are used to draw conclusions about populations.

INTRODUCTION, PAGE 2

*Watch the video for an overview of the material presented in this chapter.*
A random variable is a numerical measure of the outcome of a probability experiment. Statistics such as the sample mean, $\overline{x}$, are random variables. Statistics are random variables because the value of a statistic varies from sample to sample. For this reason, statistics have probability distributions associated with them. For example, there is a probability distribution for the sample mean and the sample proportion.

INTRODUCTION, PAGE 3

The sample mean will vary from sample to sample. Our goal in this section is to describe the distribution of the sample mean. Remember, when we describe a distribution, we do so in terms of its shape, center, and spread.

INTRODUCTION, PAGE 4

1) What is the sampling distribution of a statistic?

2) What is the sampling distribution of the sample mean $\bar{x}$ ?

3) List the three steps for determining the sampling distribution of the sample mean.

Once a particular sample is obtained, it cannot be obtained a second time.

_____

*Objective 1: Describe the Distribution of the Sample Mean: Normal Population*

OBJECTIVE 1, PAGE 1

*Answer the following after watching the video.*

4) Describe the shape of the distribution of the sample mean as the sample size increases.

OBJECTIVE 1, PAGE 1 (CONTINUED)

5) What does the mean of the distribution of the sample mean, $\bar{x}$, equal?

6) As the sample size $n$ increases, what happens to the standard deviation of the distribution of the sample mean?

OBJECTIVE 1, PAGE 2

7) List the formulas for the mean and standard deviation of the sampling distribution of $\bar{x}$.

8) What is the standard error of the mean?

OBJECTIVE 1, PAGE 3

Note, in both simulations, the standard error of the mean was close to the approximate standard error.

OBJECTIVE 1, PAGE 4

9) Describe the shape of the sampling distribution of $\bar{x}$ if the random variable $X$ is normally distributed.

OBJECTIVE 1, PAGE 7

---

**Example 1**      *Finding Probabilities of a Sample Mean*

The IQ, $X$, of humans is approximately normally distributed with mean $\mu = 100$ and standard deviation $\sigma = 15$. Compute the probability that a simple random sample of size $n = 10$ results in a sample mean greater than 110. That is, compute $P(\overline{x} > 110)$.

---

*Objective 2: Describe the Distribution of the Sample Mean: Non-normal Population*

OBJECTIVE 2, PAGE 1

*Answer the following after Activity 1: Sampling Distribution of the Sample Mean: Non-normal Population*

10) As the sample size increases, describe the effect on the center and spread of the distribution.

OBJECTIVE 2, PAGE 2

*Watch the video to help reinforce the concepts from Activity 1.*

OBJECTIVE 2, PAGE 3

11) What is the mean of the sampling distribution of the sample mean equal to? What is the standard deviation of the sampling distribution of the sample mean equal to?

12) What happens to the shape of the sampling distribution of the sample mean as the sample size increases?

13) State the Central Limit Theorem.

OBJECTIVE 2, PAGE 4

How large does the sample size need to be before we can say that the sampling distribution of $\overline{x}$ is approximately normal? The answer depends on the shape of the distribution of the underlying population. Distributions that are highly skewed will require a larger sample size for the distribution of $\overline{x}$ to become approximately normal.

OBJECTIVE 2, PAGE 5

Notice that even for a highly skewed population of household incomes for a town, the distribution of the sample mean is approximately normal for $n = 25$.

OBJECTIVE 2, PAGE 6

14) State the rule of thumb for invoking the Central Limit Theorem.

<u>OBJECTIVE 2, PAGE 9</u>

---

**Example 2**     *Weight Gain during Pregnancy*

The mean weight gain during pregnancy is 30 pounds, with a standard deviation of 12.9 pounds. Weight gain during pregnancy is skewed right. An obstetrician obtains a random sample of 35 low-income patients and determines that their mean weight gain during pregnancy was 36.2 pounds. Does this result suggest anything unusual?

---

<u>OBJECTIVE 2, PAGE 11</u>

*Watch the video for a summary of the shape, center, and spread of the distribution of the sample mean for both normal and non-normal populations.*

## Section 8.2
## Distribution of the Sample Proportion

**Objectives**

❶ Describe the Sampling Distribution of a Sample Proportion

❷ Compute Probabilities of a Sample Proportion

---

*Objective 1: Describe the Sampling Distribution of a Sample Proportion*

OBJECTIVE 1, PAGE 1

1) Define the sample proportion, $\hat{p}$.

OBJECTIVE 1, PAGE 2

| **Example 1** | *Computing a Sample Proportion* |
| --- | --- |

The Harris Poll conducted a survey of 1200 adult Americans who vacation during the summer and asked whether the individuals planned to work while on summer vacation. Of those surveyed, 552 indicated that they planned to work while on vacation. Find the sample proportion of individuals surveyed who planned to work while on summer vacation.

OBJECTIVE 1, PAGE 5

Because the value of the sample proportion, $\hat{p}$, varies from sample to sample, it is a random variable and has a probability distribution.

OBJECTIVE 1, PAGE 6

*Answer the following after watching the video.*

2) As the sample size increases, describe what happens to the shape of the sampling distribution of the sample proportion.

3) What does the mean of the sampling distribution of the sample proportion equal?

4) As the sample size increases, describe what happens to the standard deviation of the sampling distribution of the sample proportion.

OBJECTIVE 1, PAGE 7

*Answer the following after Activity 1: Sampling Distribution of the Sample Proportion.*

5) What is the mean of the distribution in all three cases?

6) What role does sample size play in the standard deviation?

OBJECTIVE 1, PAGE 7 (CONTINUED)
7) What role does sample size play in the shape of the sampling distribution of $\hat{p}$?

OBJECTIVE 1, PAGE 8
8) Under what conditions is the shape of the sampling distribution of $\hat{p}$ approximately normal?

9) State the formulas for the mean and standard deviation of the sampling distribution of $\hat{p}$.

The sample size, $n$, can be no more than 5% of the population size, $N$. That is, $n \le 0.05N$.

OBJECTIVE 1, PAGE 10

| Example 2 | *Describing the Sampling Distribution of the Sample Proportion* |
|---|---|

Based on a study conducted by the Gallup Organization, 77% of Americans believe that the state of moral values in the United States is getting worse. Suppose we obtain a simple random sample of n=60 Americans and determine which of them believe that the state of moral values in the United States is getting worse. Describe the sampling distribution of the sample proportion for Americans with this belief.

## *Objective 2: Compute Probabilities of a Sample Proportion*

<u>OBJECTIVE 2, PAGE 1</u>

Now that we know how to describe the sampling distribution of the sample proportion, we can compute probabilities involving sample proportions.

<u>OBJECTIVE 2, PAGE 2</u>

| **Example 3** | ***Computing Probabilities of a Sample Proportion*** |

According to the National Center for Health Statistics, 15% of all Americans have hearing trouble.

A) In a random sample of 120 Americans, what is the probability that at most 12% have hearing trouble?

B) Suppose that a random sample of 120 Americans who regularly listen to music using headphones results in 26 having hearing trouble. What might you conclude?

# Chapter 9 – Estimating the Value of a Parameter

**Putting It Together**

Chapters 1 through 7 laid the groundwork for the remainder of the course. These chapters dealt with data collection (Chapter 1), descriptive statistics (Chapters 2 through 4), and probability (Chapters 5 through 7).

Chapter 8 formed a bridge between probability and statistical inference by giving us models that can be used to make probability statements about the sample mean and sample proportion.

We now discuss inferential statistics—the process of generalizing information obtained from a sample to a population. We will study two areas of inferential statistics:

Estimation: Sample data are used to estimate the value of unknown parameters such as $\mu$ or $p$.

Hypothesis testing: Statements regarding a characteristic of one or more populations are tested using sample data.

In this chapter, we discuss estimation of an unknown parameter, and in the next chapter we discuss hypothesis testing.

## Section 9.1
## Estimating a Population Proportion

**Objectives**

&#10122; Obtain a Point Estimate for the Population Proportion

&#10123; Construct and Interpret a Confidence Interval for the Population Proportion

&#10124; Determine the Sample Size Necessary for Estimating a Population Proportion within a Specified Margin of Error

---

### *Objective 1: Obtain a Point Estimate for the Population Proportion*

OBJECTIVE 1, PAGE 1

1) State the definition of a point estimate.

OBJECTIVE 1, PAGE 2

| **Example 1** | *Obtaining a Point Estimate of a Population Proportion* |
| --- | --- |

The Gallup Organization conducted a poll in April 2017 in which a simple random sample of 1019 Americans aged 18 and older were asked, "Do you regard the income tax that you will have to pay this year as fair?" Of the 1019 adult Americans surveyed, 620 said yes. Obtain a point estimate for the proportion of Americans aged 18 and older who believe that the amount of income tax they pay is fair.

---

### *Objective 2: Construct and Interpret a Confidence Interval for the Population Proportion*

OBJECTIVE 2, PAGE 1

Statistics such as $\hat{p}$ vary from sample to sample.

Due to variability in the sample proportion, we report a range (or *interval*) of values, including a measure of the likelihood that the interval includes the unknown population parameter.

OBJECTIVE 2, PAGE 2

*Watch the video, which discusses the logic behind the construction of confidence intervals for a population proportion. Answer the following as you watch.*

2) Give the definition for a confidence interval for an unknown parameter.

3) What does the level of confidence represent?

4) What is the form of confidence interval estimates for a population parameter?

**Note: Review of the Sampling Distribution of the Sample Proportion**

- The shape of the distribution of all possible sample proportions is approximately normal provided $np(1-p) \geq 10$ and the sample size is no more than 5% of the population size. That is, $n \leq 0.05N$.

- The mean of the distribution of the sample proportions equals the population proportion. That is, $\mu_{\hat{p}} = p$.

- The standard deviation of the distribution of the sample proportion (the standard error) is

$$\sigma_{\hat{p}} = \frac{\sigma}{\sqrt{n}}$$

5) How is the margin of error computed for a 95% confidence interval for a population proportion?

- For a 95% confidence interval, any sample proportion that lies within 1.96 standard errors of the population proportion will result in a confidence interval that includes $p$. This will happen in 95% of all possible samples.

- Any sample proportion that is more than 1.96 standard errors from the population proportion will result in a confidence interval that does not contain $p$. This will happen in 5% of all possible samples (those sample proportions in the tails of the distribution).

- Whether a confidence interval contains the population parameter depends solely on the value of the sample statistic.

- Any sample statistic that is in the tails of the sampling distribution will result in a confidence interval that does not include the population parameter.

Objective 2, Page 3

**Key Ideas Regarding Confidence Intervals**

- A confidence interval for an unknown parameter consists of an interval of numbers based on a point estimate.

- The level of confidence represents the expected proportion of intervals that will contain the parameter if a large number of different samples is obtained. The level of confidence is denoted $(1 - \alpha) \cdot 100\%$.

- Whether a confidence interval contains the population parameter depends solely on the value of the sample statistic. Any sample statistic that is in the tails of the sampling distribution will result in a confidence interval that does not include the population parameter.

Objective 2, Page 6

*Answer the following after Activity 1: Illustrating the Meaning of Level of Confidence Using Simulation*

6) What proportion of the 95% confidence intervals generated by simulation would you expect to contain the population parameter $p$? What proportion of the 99% confidence intervals would you expect to contain the population parameter $p$?

Objective 2, Page 7

*Watch the video to reinforce the concepts learned from Activity 1 (Objective 2, Page 6).*

OBJECTIVE 2, PAGE 10

*Answer the following after watching the video.*

7) What does the "95%" in a 95% confidence interval represent?

OBJECTIVE 2, PAGE 12

$(1-\alpha)\cdot 100\%$ of all sample proportions will result in confidence intervals that contain the population proportion. The sample proportions that are in the tails of the distribution, outside the interval

$$\left( p - z_{\frac{\alpha}{2}} \cdot \sigma_{\hat{p}},\ p + z_{\frac{\alpha}{2}} \cdot \sigma_{\hat{p}} \right),$$ will result in confidence intervals that contain the population proportion.

OBJECTIVE 2, PAGE 13

8) What does the critical value of a distribution represent?

9) As the level of confidence increases, what happens to the critical value?

10) List the critical value associated with the given level of confidence.

A) 90%                    B) 95%                    C) 99%

OBJECTIVE 2, PAGE 14

11) State the interpretation of a confidence interval.

OBJECTIVE 2, PAGE 15

---

**Example 2**     *Interpreting a Confidence Interval*

The Gallup Organization conducted a poll in April 2017 in which a simple random sample of 1019 Americans aged 18 and older were asked, "Do you regard the income tax that you will have to pay this year as fair?" We learned from Example 1 that the proportion of those surveyed who responded yes was 0.608. Gallup reported its "survey methodology" as follows:

Results are based on telephone interviews with a random sample of 1019 national adults, aged 18 and older. For results based on the total sample of national adults, one can say with 95% confidence that the maximum margin of sampling error is 4 percentage points.

Determine and interpret the confidence interval for the proportion of Americans aged 18 and older who believe the amount of federal income tax they have to pay is fair.

---

OBJECTIVE 2, PAGE 17

**Note:**

A 90% level of confidence does not tell us that there is a 90% probability that the parameter lies between the lower and upper bound.

It means that the interval includes the unknown parameter for 90% of all samples.

12) Explain the method for constructing a confidence interval about the population proportion, $p$.

**Note:** It must be the case that $n\hat{p}(1-\hat{p}) \geq 10$ and $n \leq 0.05N$ to construct this interval. Use $\hat{p}$ in place of $p$ in the standard deviation. This is because $p$ is unknown, and $\hat{p}$ is the best point estimate of $p$.

| **Example 3** | ***Constructing a Confidence Interval for a Population Proportion*** |
|---|---|

In the Parent-Teen Cell Phone Survey conducted by Princeton Survey Research Associates International, 800 randomly sampled 16- to 17-year-olds living in the United States were asked whether they have ever used their cell phone to text while driving. Of the 800 teenagers surveyed, 272 indicated that they text while driving. Obtain a 95% confidence interval for the proportion of 16- to 17-year-olds who text while driving.

13) State the formula for the margin of error, $E$, for a $(1-\alpha)\cdot 100\%$ confidence interval for a population proportion.

| **Example 4** | *The Role of the Level of Confidence in the Margin of Error* |
|---|---|

In the Parent-Teen Cell Phone Survey conducted by Princeton Survey Research Associates International, 800 randomly sampled 16- to 17-year-olds living in the United States were asked whether they have ever used their cell phone to text while driving. Of the 800 teenagers surveyed, 272 indicated that they text while driving. From the last example, we concluded that we are 95% confident that the proportion of 16- to 17-year olds who text while driving is between 0.307 and 0.373. Determine the effect on the margin of error by increasing the level of confidence from 95% to 99%.

14) As the sample size, $n$, increases, what happens to the margin of error?

OBJECTIVE 2, PAGE 24

*Answer the following after watching the video.*

15) If the normality condition is not satisfied, how does the proportion of intervals that capture the parameter compare to the level of confidence?

---

### *Objective 3: Determine the Sample Size Necessary for Estimating a Population Proportion within a Specified Margin of Error*

OBJECTIVE 3, PAGE 1

*Watch the video for an explanation of where the formulas for determining sample size for estimating a population proportion within a specified margin of error come from.*

OBJECTIVE 3, PAGE 2

16) List the formula for the sample size required to obtain a $(1-\alpha)\cdot 100\%$ confidence interval for $p$ with a margin of error $E$, if $\hat{p}$ is a prior estimate of $p$.

17) List the formula for the sample size required to obtain a $(1-\alpha)\cdot 100\%$ confidence interval for $p$ with a margin of error $E$, if a prior estimate of $p$ is unavailable.

**Example 5**     *Determining Sample Size*

An economist wants to know if the proportion of the U.S. population who commutes to work via car-pooling is on the rise. What size sample should be obtained if the economist wants an estimate within 2 percentage points of the true proportion with 90% confidence?

A) Assume that the economist uses the estimate of 10% obtained from the American Community Survey.

B) Assume that the economist does not use any prior estimates.

# Section 9.2
# Estimating a Population Mean

**Objectives**

❶ Obtain a Point Estimate for the Population Mean

❷ State Properties of Student's *t*-Distribution

❸ Determine *t*-Values

❹ Construct and Interpret a Confidence Interval for a Population Mean

❺ Determine the Sample Size Necessary for Estimating a Population Mean within a Given Margin of Error

---

## Objective 1: Obtain a Point Estimate for the Population Mean

Objective 1, Page 1

1) What is the point estimate for a population mean $\mu$?

Objective 1, Page 2

| **Example 1** | ***Computing a Point Estimate of the Population Mean*** |
|---|---|

The website fueleconomy.gov allows drivers to report the miles per gallon of their vehicle. The data in Table 2 show the reported miles per gallon of 2011 Ford Focus automobiles for 16 different owners. Obtain a point estimate of the population mean miles per gallon of a 2011 Ford Focus.

**Table 2**

| | | | |
|---|---|---|---|
| 35.7 | 37.2 | 34.1 | 38.9 |
| 32.0 | 41.3 | 32.5 | 37.1 |
| 37.3 | 38.8 | 38.2 | 39.6 |
| 32.2 | 40.9 | 37.0 | 36.0 |

## Objective 2: State Properties of Student's t-Distribution

<u>OBJECTIVE 2, PAGE 1</u>

Recall that the distribution of $\bar{x}$ is approximately normal if the population from which the sample is drawn is normal or the sample size is sufficiently large. In addition, the distribution of $\bar{x}$ has the same mean as the parent population, $\mu_{\bar{x}} = \mu$, and a standard deviation equal to the parent population's standard deviation divided by the square root of the sample size, $\sigma_{\bar{x}} = \dfrac{\sigma}{\sqrt{n}}$.

<u>OBJECTIVE 2, PAGE 2</u>

Using $\bar{x} \pm z_{\frac{\alpha}{2}} \cdot \dfrac{\sigma}{\sqrt{n}}$ for a confidence interval for the mean presents a problem because it is not likely that we know the population standard deviation $(\sigma)$ but not know the population mean $(\mu)$.

<u>OBJECTIVE 2, PAGE 3</u>

Using $s$ as an estimate for $\sigma$ also presents a problem because the sample standard deviation, $s$, is a statistic and therefore will vary from sample to sample. Using the normal model to determine the critical value, $z_{\frac{\alpha}{2}}$, in the margin of error does not take into account the additional variability introduced by using $s$ in place of $\sigma$. A new model must be used to determine the margin of error in a confidence interval that accounts for the additional variability. This leads to the story of William Gosset.

<u>OBJECTIVE 2, PAGE 4</u>

2) What was the name of the brewery that Gosset worked for? What pseudonym did he choose to publish his results about a model that accounts for the additional variability introduced by using $s$ in place of $\sigma$ when determining margin of error?

<u>OBJECTIVE 2, PAGE 5</u>

*Watch the video that uses simulation to illustrate some of the work Gosset did by hand to develop his sampling distribution—Student's t-distribution..*

OBJECTIVE 2, PAGE 6

3) Suppose that a simple random sample of size $n$ is taken from a population. If the population from which the sample is drawn follows a normal distribution, what does the distribution of $t = \dfrac{\bar{x} - \mu}{\frac{s}{\sqrt{n}}}$ follow?

OBJECTIVE 2, PAGE 7

4) State six properties of the $t$-distribution.

---

## Objective 3: Determine t-Values

OBJECTIVE 3, PAGE 1

5) What does $t_\alpha$ represent?

**Example 2**   *Finding t-Values*

Find the *t*-value such that the area under the *t*-distribution to the right of the *t*-value is 0.10, assuming 15 degrees of freedom (df). That is, find $t_{0.10}$ with 15 degrees of freedom.

OBJECTIVE 3, PAGE 4

If the degrees of freedom we desire are not listed in Table VII, choose the closest number in the "df" column.

In addition, the last row of Table VII lists the *z*-values from the standard normal distribution. Use these values when the degrees of freedom are more than 1000 because the *t*-distribution starts to behave like the standard normal distribution as *n* increases.

*Objective 4: Construct and Interpret a Confidence Interval for a Population Mean*

OBJECTIVE 4, PAGE 1

6) List the three conditions required for constructing a confidence interval for a population mean $\mu$.

7) List the formulas for the lower bound and upper bound for a $(1-\alpha)\cdot 100\%$ confidence interval for the population mean, $\mu$.

OBJECTIVE 4, PAGE 2

8) What does it mean when we say that the procedure for constructing a confidence interval is robust?

9) Because the sample mean and sample standard deviation are not resistant to outliers, sample data should always be inspected for serious departures from normality and for outliers. What tools can be used to check for serious departures from normality and for outliers?

OBJECTIVE 4, PAGE 3

---

**Example 3**     *Constructing a Confidence Interval about a Population Mean*

The website fueleconomy.gov allows drivers to report the miles per gallon of their vehicle. The data in Table 3 show the reported miles per gallon of 2011 Ford Focus automobiles for 16 different owners. Treat the sample as a simple random sample of all 2011 Ford Focus automobiles. Construct a 95% confidence interval for the mean miles per gallon of a 2011 Ford Focus. Interpret the interval.

**Table 3**

| | | | |
|------|------|------|------|
| 35.7 | 37.2 | 34.1 | 38.9 |
| 32.0 | 41.3 | 32.5 | 37.1 |
| 37.3 | 38.8 | 38.2 | 39.6 |
| 32.2 | 40.9 | 37.0 | 36.0 |

---

OBJECTIVE 4, PAGE 4

The *t*-distribution gives a larger critical value than the *z*-distribution, so the width of the confidence interval is wider when it is constructed using Student's *t*-distribution.

OBJECTIVE 4, PAGE 7

*Answer the following after Activity 1: When Model Requirements Fail.*

10) What happens to the proportion of intervals that capture the population mean as the sample size increases?

OBJECTIVE 4, PAGE 9

If the requirements to compute a *t*-interval are not met, one option is to use resampling methods, such as bootstrapping. Bootstrapping is presented later in this chapter.

---

## Objective 5: Determine the Sample Size Necessary for Estimating a Population Mean within a Given Margin of Error

OBJECTIVE 5, PAGE 1

11) State the formula for margin of error when constructing a confidence interval about the population mean.

12) List the formula for the sample size required to obtain a $(1-\alpha)\cdot 100\%$ confidence interval for $\mu$ with a margin of error $E$.

OBJECTIVE 5, PAGE 2

| **Example 4** | **Determining Sample Size** |
| --- | --- |

We again consider the problem of estimating the miles per gallon of a 2011 Ford Focus. How large a sample is required to estimate the mean miles per gallon within 0.5 miles per gallon with 95% confidence? Note: The sample standard deviation is s=2.92 mpg.

## Section 9.3
## Putting It Together: Which Procedure Do I Use?

**Objectives**

❶ Determine the Appropriate Confidence Interval to Construct

---

***Objective 1: Determine the Appropriate Confidence Interval to Construct***

OBJECTIVE 1, PAGE 1

*Answer the following after watching the video.*

1) What type of data are needed to construct a confidence interval for a population proportion, $p$?

2) Besides the fact that the sample must be obtained by simple random sampling or through a randomized experiment, list the two conditions that must be met when constructing a confidence interval for a population proportion, $p$.

3) What type of data are needed to construct a confidence interval for a population mean, $\mu$?

4) Besides the facts that the sample must be obtained by simple random sampling or through a randomized experiment and that the sample size must be small relative to the size of the population, what other condition must be satisfied?

**Flowchart for Determining Which Type of Confidence to Construct**

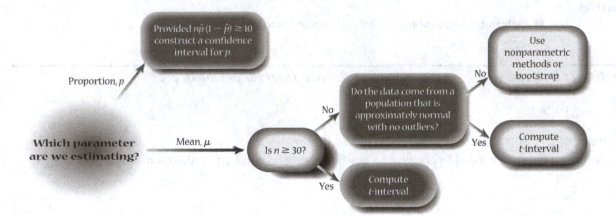

## Section 9.4
## Estimating with Bootstrapping

**Objectives**

❶ Estimate a Parameter Using the Bootstrap Method

---

*Objective 1: Estimate a Parameter Using the Bootstrap Method*

OBJECTIVE 1, PAGE 1

1) State the definition for bootstrapping.

OBJECTIVE 1, PAGE 2

Watch the video to learn about the logic behind the bootstrap method.

OBJECTIVE 1, PAGE 3

2) List the two basic requirements that must be satisfied to use bootstrapping.

3) What two percentiles are associated with a 95% confidence interval?

OBJECTIVE 1, PAGE 4

4) List the three steps of the bootstrap algorithm.

OBJECTIVE 1, PAGE 7

**Example 1**     *Using the Bootstrap Method to Construct a 95% Confidence Interval*

The website fueleconomy.gov allows drivers to report the miles per gallon of their vehicle. The data in Table 4 show the reported miles per gallon of 2011 Ford Focus automobiles for 16 different owners. Treat the sample as a simple random sample of all 2011 Ford Focus automobiles. Construct a 95% confidence interval for the mean miles per gallon of a 2011 Ford Focus using a bootstrap sample. Interpret the interval.

**Table 4**

| | | | |
|------|------|------|------|
| 35.7 | 37.2 | 34.1 | 38.9 |
| 32.0 | 41.3 | 32.5 | 37.1 |
| 37.3 | 38.8 | 38.2 | 39.6 |
| 32.2 | 40.9 | 37.0 | 36.0 |

OBJECTIVE 1, PAGE 10

The confidence interval constructed using the bootstrap method is slightly different than the confidence interval using Student's *t*-distribution, but the two intervals are very similar.

Because bootstrapping relies on randomization, results will vary each time the method is used. So, the results you obtain by following Example 1 may differ from the results we obtained, but your results should be close.

OBJECTIVE 1, PAGE 12

Many statistical spreadsheets (such as StatCrunch) have built-in algorithms that do bootstrapping.

OBJECTIVE 1, PAGE 13

---

**Example 2**     *Using StatCrunch's Resample Command to Obtain a Bootstrap Confidence Interval*

Use StatCrunch to estimate a 95% confidence interval for the mean miles per gallon of a 2011 Ford Focus based on the sample data in Table 4.

**Table 4**

| | | | |
|------|------|------|------|
| 35.7 | 37.2 | 34.1 | 38.9 |
| 32.0 | 41.3 | 32.5 | 37.1 |
| 37.3 | 38.8 | 38.2 | 39.6 |
| 32.2 | 40.9 | 37.0 | 36.0 |

---

OBJECTIVE 1, PAGE 15

To construct a bootstrap confidence interval for a proportion, we need raw data using 0 for a failure and 1 for a success. The sample proportion is then estimated using the mean of the 0s and 1s.

OBJECTIVE 1, PAGE 16

5) List three items to be aware of when constructing confidence intervals using the bootstrap approach.

# Chapter 10 – Hypothesis Tests Regarding a Parameter

## Putting It Together

In Chapter 9, we mentioned there are two types of inferential statistics:

(1) Estimation
(2) Hypothesis testing

We have already discussed procedures for estimating the population proportion and the population mean.

We now focus our attention on hypothesis testing. Hypothesis testing is used to test statements regarding a characteristic of one or more populations. In this chapter, we will test hypotheses regarding a single population parameter, including the population proportion and the population mean.

## Section 10.1
## The Language of Hypothesis Testing

**Objectives**

❶ Determine the Null and Alternative Hypotheses

❷ Explain Type I and Type II Errors

❸ State Conclusions to Hypothesis Tests

<u>INTRODUCTION, PAGE 2</u>

*Answer the following after watching the video.*

1) After determining that the first five tosses are all tails is unlikely, what are the two possible conclusions that can be drawn?

---

### *Objective 1: Determine the Null and Alternative Hypotheses*

<u>OBJECTIVE 1, PAGE 1</u>

*Answer the following (2–8) after watching the video.*

2) What is a hypothesis?

3) Why do we test statements about a population parameter using sample data?

4) State the definition of hypothesis testing.

<u>OBJECTIVE 1, PAGE 1 (CONTINUED)</u>
5) List the 3 steps in hypothesis testing.

6) State the definition of the null hypothesis.

7) State the definition of the alternative hypothesis.

8) List the three ways to set up the null and alternative hypotheses.
Two-tailed test

Left-tailed test

Right-tailed test

9) What type of tests are referred to as one-tailed tests?

10) What determines the structure of the alternative hypothesis (two-tailed, left-tailed, or right-tailed?)

---

**Example 1**    *Forming Hypotheses*

For each situation, determine the null and alternative hypotheses. State whether the test is two-tailed, left-tailed, or right-tailed.

A) The Medco pharmaceutical company has just developed a new antibiotic for children. Two percent of children taking competing antibiotics experience headaches as a side effect. A researcher for the Food and Drug Administration wants to know if the percentage of children taking the new antibiotic and experiencing headaches as a side effect is more than 2%.

B) The Blue Book value of a used three-year-old Chevy Corvette Z06 is $56,130. Grant wonders if the mean price of a used three-year-old Chevy Corvette Z06 in the Miami metropolitan area is different from $56,130.

C) The standard deviation of the contents in a 64-ounce bottle of detergent using an old filling machine is 0.23 ounce. The manufacturer wants to know if a new filling machine has less variability.

---

### Objective 2: Explain Type I and Type II Errors

11) What type of error is called a Type I error?

12) What type of error is called a Type II error?

*Answer the following after watching the video.*

13) In a jury trial, what are the null and alternative hypotheses?

14) What jury decision is associated with rejecting the null hypothesis?

15) What jury decision is associated with failing to reject the null hypothesis?

16) Is the null hypothesis ever declared "true"?

17) In a jury trial, what decision is equivalent to making a Type I error?

18) In a jury trial, what decision is equivalent to making a Type II error?

OBJECTIVE 2, PAGE 2 (CONTINUED)
19) Sketch the chart that illustrates the four outcomes from hypothesis testing.

OBJECTIVE 2, PAGE 3

**Example 2**    *Type I and Type II Errors*

The Medco pharmaceutical company has just developed a new antibiotic. Two percent of children taking competing antibiotics experience headaches as a side effect. A researcher for the Food and Drug Administration wishes to know if the percentage of children taking the new antibiotic who experience a headache as a side effect is more than 2%.

The researcher conducts a hypothesis test with $H_0 : p = 0.02$ and $H_1 : p > 0.02$.

Explain what it would mean to make a (A) Type I error and (B) Type II error.

OBJECTIVE 2, PAGE 5
20) What symbols do we use to denote the probability of making a Type I error and the probability of making a Type II error?

<u>OBJECTIVE 2, PAGE 6</u>
21) What does the level of significance represent?




22) What does the choice of the level of significance depend on?




23) Why is the level of significance not always set at $\alpha = 0.01$?

---

## *Objective 3: State Conclusions to Hypothesis Tests*

<u>OBJECTIVE 3, PAGE 1</u>
It is important to recognize that we never accept the null hypothesis. Sample evidence can never prove the null hypothesis to be true. By not rejecting the null hypothesis, we are saying that the evidence indicates that the null hypothesis could be true or that the sample evidence is consistent with the statement in the null hypothesis.

**Example 3**     *Stating the Conclusion*

The Medco pharmaceutical company has just developed a new antibiotic. Two percent of children taking competing antibiotics experience a headache as a side effect. A researcher for the Food and Drug Administration believes that the proportion of children taking the new antibiotic who experience a headache as a side effect is more than 0.02. So the null hypothesis is $H_0 : p = 0.02$ and the alternative hypothesis is $H_1 : p > 0.02$.

A) Suppose the sample evidence indicates that the null hypothesis is rejected. State the conclusion.

B) Suppose the sample evidence indicates that the null hypothesis is not rejected. State the conclusion.

## Section 10.2
## Hypothesis Tests for a Population Proportion

**Objectives**

❶ Explain the Logic of Hypothesis Testing

❷ Test Hypotheses about a Population Proportion

❸ Test Hypotheses about a Population Proportion Using the Binomial Probability Distribution

**Note:** Your instructor may cover either Section 10.2 or Sections 10.2A and 10.2B. Be sure you know which sections are part of your course.

### Objective 1: Explain the Logic of Hypothesis Testing

OBJECTIVE 1, PAGE 1

The applet on page 2 will help you to determine what would be convincing evidence that the population proportion of registered voters who are in favor of a certain policy is greater than 50%.

OBJECTIVE 1, PAGE 2

*Answer the following after using the Political Poll applet.*

1) What is the center of the distribution when you simulate 1000 samples of 500 registered voters?

2) As the sample proportion increases from 52% to 54% to 56%, what happens to the proportion of surveys that resulted in a sample proportion that was greater than or equal to the given sample proportion?

3) Explain how you determined whether the proportion of voters in favor of this policy is greater than 0.5.

OBJECTIVE 1, PAGE 6

4) Give the definition of what it means for a result to be statistically significant.

<u>OBJECTIVE 1, PAGE 7</u>
**Note:** The sample distribution of $\hat{p}$ is approximately normal, with mean $\mu_{\hat{p}} = p$ and standard deviation

$\sigma_{\hat{p}} = \sqrt{\dfrac{p(1-p)}{n}}$ , provided that the following requirements are satisfied:

The sample is a simple random sample.

$np(1-p) \geq 10$

The sampled values are independent of each other $(n \leq 0.05N)$.

<u>OBJECTIVE 1, PAGE 8</u>
A criterion for testing hypotheses is to determine how likely the observed sample proportion is under the assumption that the statement in the null hypothesis is true.

For example, for the scenario in Part (C) of the Logic of Hypothesis Testing Activity on page 2, the probability of obtaining a sample proportion of 0.52 or higher from a population whose proportion is assumed to be $p = 0.5$ is 0.1855.

<u>OBJECTIVE 1, PAGE 9</u>
The likelihood of obtaining a sample statistic can be obtained either through simulation or through the use of the normal model. Both approaches give similar results.

<u>OBJECTIVE 1, PAGE 10</u>
5) Give the definition of a *P*-value.

6) Explain how to determine whether the null hypothesis should be rejected using the *P*-value approach.

<u>OBJECTIVE 1, PAGE 11</u>
Figure 4 illustrates that obtaining a sample proportion of 0.54 or higher from a population whose proportion is 0.5 is unlikely. Therefore, we reject the null hypothesis that $p = 0.5$ and conclude that $p > 0.5$. We do not know what the population proportion of registered voters who are in favor of the policy is, but we have evidence to say that it is greater than 0.5 (a majority).

## Objective 2: Test Hypotheses about a Population Proportion

<u>OBJECTIVE 2, PAGE 1</u>

7) What are the three conditions that must be satisfied before testing a hypothesis regarding a population proportion, $p$?

8) State the five steps for testing a hypothesis about a population proportion, $p$.

Step 1

Step 2

Step 3 (By Hand)                    Step 3 (Using Technology)

Step 4

Step 5

**Example 1**    *Testing a Hypothesis about a Population Proportion: Left-Tailed Test*

The two major college entrance exams that a majority of colleges accept for student admission are the SAT and ACT. ACT looked at historical records and established 22 as the minimum ACT math score for a student to be considered prepared for college mathematics. (Note: "Being prepared" means that there is a 75% probability of successfully completing College Algebra in college.) An official with the Illinois State Department of Education wonders whether less than half of the students in her state are prepared for College Algebra. She obtains a simple random sample of 500 records of students who have taken the ACT and finds that 219 are prepared for college mathematics (that is, scored at least 22 on the ACT math test). Does this represent significant evidence that less than half of Illinois students are prepared for college mathematics upon graduation from a high school? Use the $\alpha = 0.05$ level of significance. Data from ACT High School Profile Report.

| Example 2 | *Testing a Hypothesis about a Population Proportion: Two-Tailed Test* |

When asked the following question, "Which do you think is more important—protecting the right of Americans to own guns or controlling gun ownership?" 46% of Americans said that protecting the right to own guns is more important. The Pew Research Center surveyed 1267 randomly selected Americans with at least a bachelor's degree and found that 559 believed that protecting the right to own guns is more important. Does this result suggest that the proportion of Americans with at least a bachelor's degree feel differently than the general American population when it comes to gun control? Use the $\alpha = 0.1$ level of significance.

9) Explain how to make a decision about the null hypothesis when performing a two-tailed test using confidence intervals.

| Example 3 | *Testing a Hypothesis Using a Confidence Interval* |

A 2009 study by Princeton Survey Research Associates International found that 34% of teenagers text while driving. A recent study conducted by AT&T found that 515 of 1200 randomly selected teens had texted while driving. Do the results of this study suggest that the proportion of teens who text while driving has changed since 2009? Use a 95% confidence interval to answer the question.

## Objective 3: Test Hypotheses about a Population Proportion Using the Binomial Probability Distribution

OBJECTIVE 3, PAGE 1

For the sampling distribution of $\hat{p}$ to be approximately normal, we require that $np(1-p)$ be at least 10. If this requirement is not satisfied we use the binomial probability formula to determine the $P$-value.

OBJECTIVE 3, PAGE 2

**Example 4**  *Hypothesis Test for a Population Proportion: Small Sample Size*

According to the U.S. Department of Agriculture, 48.9% of males aged 20 to 39 years consume the recommended daily requirement of calcium. After an aggressive "Got Milk" advertising campaign, the USDA conducts a survey of 35 randomly selected males aged 20 to 39 and finds that 21 of them consume the recommended daily allowance (RDA) of calcium. At the $\alpha = 0.10$ level of significance, is there evidence to conclude that the percentage of males aged 20 to 39 who consume the RDA of calcium has increased?

## Section 10.2A
## Hypothesis Tests on a Population Proportion with Simulation

**Objectives**

❶ Explain the Logic of the Simulation Method

❷ Test Hypotheses about a Population Proportion Using the Simulation Method

---

**Note:** Your instructor may cover either Section 10.2 or Sections 10.2A and 10.2B. Be sure you know which sections are part of your course.

### *Objective 1: Explain the Logic of the Simulation Method*

OBJECTIVE 1, PAGE 1

1) Give the definition of what it means for a result to be statistically significant.

OBJECTIVE 1, PAGE 3

To determine if sample results are statistically significant, we build a model that generates data randomly under the assumption the statement in the null hypothesis is true. Call this model the **null model**. Then compare the results of the randomly generated data from the null model to those observed to see if the observed results are unusual.

OBJECTIVE 1, PAGE 4

Repeating the simulation could lead to different results because of the randomness of the process.

OBJECTIVE 1, PAGE 5

If we repeat the process of simulation many, many times, we will be able to build a null model and use it to determine how often results such as those observed occur in the random process.

OBJECTIVE 1, PAGE 6

2) Give the definition of a *P*-value.

OBJECTIVE 1, PAGE 7

In the ESP study, we used the number of heads observed, 24, as the **test statistic**. We then determined the proportion of times we observed 24 or more heads in many repetitions using the null model.

Instead of using the number of heads as the test statistic, we could have used the sample proportion,

$\hat{p} = \dfrac{24}{40} = 0.6,$ as the test statistic. We would then determine the proportion of times we observed a

proportion of heads of 0.6 or higher in many repetitions using the null model.

OBJECTIVE 1, PAGE 10

Answer the following after watching the video.

3) What is the variable of interest in this study?

4) Why does it make sense to analyze this problem using proportions?

5) State the null and alternative hypotheses for this problem.

6) What is the sample proportion for this problem?

We are trying to determine how likely is it to obtain a sample proportion of 0.367 or lower from a population whose proportion is 0.42.

OBJECTIVE 1, PAGE 11

Watch the video to see how to use the urn applet in StatCrunch to simulate the results.

7) List the two guidelines for determining whether to use the Coin Flipping applet or the Urn applet for simulation.

8) Explain how to determine whether the null hypothesis should be rejected using the *P*-value approach.

9) List the rule of thumb for determining whether the null hypothesis should be rejected.

---

## *Objective 2: Test Hypotheses about a Population Proportion Using the Simulation Method*

10) State the five steps for testing a hypothesis about a population proportion using simulation.

Step 1

Step 2

Step 3

Step 4

Step 5

| Example 1 | *Testing a Hypothesis about a Population Proportion: Left-Tailed Test* |

The two major college entrance exams that a majority of colleges accept for student admission are the SAT and ACT. ACT looked at historical records and established 22 as the minimum ACT math score for a student to be considered prepared for college mathematics. (Note: "Being prepared" means that there is a 75% probability of successfully completing College Algebra in college.) An official with the Illinois State Department of Education wonders whether less than half of the students in her state are prepared for College Algebra. She obtains a simple random sample of 500 records of students who have taken the ACT and finds that 219 are prepared for college mathematics (that is, scored at least 22 on the ACT math test). Does this represent significant evidence that less than half of Illinois students are prepared for college mathematics upon graduation from a high school?

| **Example 2** | ***Testing a Hypothesis about a Population Proportion: Two-Tailed Test*** |

When asked the following question, "Which do you think is more important—protecting the right of Americans to own guns or controlling gun ownership?" 46% of Americans said that protecting the right to own guns is more important. The Pew Research Center surveyed 1267 randomly selected Americans with at least a bachelor's degree and found that 559 believed that protecting the right to own guns is more important. Does this result suggest that the proportion of Americans with at least a bachelor's degree feel differently than the general American population when it comes to gun control?

## Section 10.2B
## Hypothesis Tests on a Population Proportion Using the Normal Model

**Objectives**

❶ Explain the Logic of Hypothesis Testing Using the Normal Model

❷ Test Hypotheses about a Population Proportion Using the Normal Model

❸ Test Hypotheses about a Population Proportion Using the Binomial Probability Distribution

---

**Note:** Your instructor may cover either Section 10.2 or Sections 10.2A and 10.2B. Be sure you know which sections are part of your course.

---

### *Objective 1: Explain the Logic of Hypothesis Testing Using the Normal Model*

OBJECTIVE 1, PAGE 1

One of the examples presented in Section 10.2A dealt with congressional districts. Recall the scenario. Prior to redistricting, the proportion of registered Republicans in a congressional district was 0.42. After redistricting, a random sample of 60 voters resulted in 22 being Republican. The goal of the research was to determine if the proportion of voters in the district registered as Republican decreased after redistricting. The hypotheses to be tested were $H_0 : p = 0.42$ versus $H_1 : p < 0.42$, and the StatCrunch urn applet was used.

*Watch the video to continue the discussion of the problem, then answer the following.*

1) What is the shape of the sampling distribution of the sample proportions?

2) List the formulas for the mean and standard deviation of the sampling distribution of the sample proportion and use them to compute the theoretical mean and standard deviation. How close were they to the mean and standard deviation of the 5000 simulated sample proportions?

3) Verify that the normal model may be used to describe the sampling distribution of the sample proportion.

4) Use the normal model to determine the probability of observing 22 or fewer Republicans in a district whose population proportion is 0.42.

OBJECTIVE 1, PAGE 2

The sample distribution of $\hat{p}$ is approximately normal, with mean $\mu_{\hat{p}} = p$ and standard deviation

$\sigma_{\hat{p}} = \sqrt{\dfrac{p(1-p)}{n}}$ , provided that the following requirements are satisfied:

The sample is a simple random sample.

$np(1-p) \geq 10$

The sampled values are independent of each other $(n \leq 0.05N)$.

Rather than using simulation to approximate a $P$-value, we can use a normal model to determine the $P$-value for a hypothesis test about a population proportion.

## Objective 2: Test Hypotheses about a Population Proportion Using the Normal Model

OBJECTIVE 2, PAGE 1

5) What are the three conditions that must be satisfied before testing a hypothesis regarding a population proportion, $p$?

6) State the five steps for testing a hypothesis about a population proportion, $p$.

Step 1

Step 2

Step 3 (By Hand)                                    Step 3 (Using Technology)

Step 4

Step 5

| Example 1 | Testing a Hypothesis about a Population Proportion: Left-Tailed Test |
|---|---|

The two major college entrance exams that a majority of colleges accept for student admission are the SAT and ACT. ACT looked at historical records and established 22 as the minimum ACT math score for a student to be considered prepared for college mathematics. (Note: "Being prepared" means that there is a 75% probability of successfully completing College Algebra in college.) An official with the Illinois State Department of Education wonders whether less than half of the students in her state are prepared for College Algebra. She obtains a simple random sample of 500 records of students who have taken the ACT and finds that 219 are prepared for college mathematics (that is, scored at least 22 on the ACT math test). Does this represent significant evidence that less than half of Illinois students are prepared for college mathematics upon graduation from a high school? Use the $\alpha = 0.05$ level of significance. Data from ACT High School Profile Report.

| Example 2 | *Testing a Hypothesis about a Population Proportion: Two-Tailed Test* |

When asked the following question, "Which do you think is more important—protecting the right of Americans to own guns or controlling gun ownership?", 46% of Americans said that protecting the right to own guns is more important. The Pew Research Center surveyed 1267 randomly selected Americans with at least a bachelor's degree and found that 559 believed that protecting the right to own guns is more important. Does this result suggest that the proportion of Americans with at least a bachelor's degree feel differently than the general American population when it comes to gun control? Use the $\alpha = 0.1$ level of significance.

7) Explain how to make a decision about the null hypothesis when performing a two-tailed test using confidence intervals.

| **Example 3** | ***Testing a Hypothesis Using a Confidence Interval*** |

A 2009 study by Princeton Survey Research Associates International found that 34% of teenagers text while driving. A recent study conducted by AT&T found that 515 of 1200 randomly selected teens had texted while driving. Do the results of this study suggest that the proportion of teens who text while driving has changed since 2009? Use a 95% confidence interval to answer the question.

## Objective 3: Test Hypotheses about a Population Proportion Using the Binomial Probability Distribution

OBJECTIVE 3, PAGE 1

For the sampling distribution of $\hat{p}$ to be approximately normal, we require that $np(1-p)$ be at least 10. If this requirement is not satisfied we use the binomial probability formula to determine the $P$-value.

OBJECTIVE 3, PAGE 2

**Example 4** *Hypothesis Test for a Population Proportion: Small Sample Size*

According to the U.S. Department of Agriculture, 48.9% of males aged 20 to 39 years consume the recommended daily requirement of calcium. After an aggressive "Got Milk" advertising campaign, the USDA conducts a survey of 35 randomly selected males aged 20 to 39 and finds that 21 of them consume the recommended daily allowance (RDA) of calcium. At the α = 0.10 level of significance, is there evidence to conclude that the percentage of males aged 20 to 39 who consume the RDA of calcium has increased?

## Section 10.3
## Hypothesis Tests for a Population Mean

**Objectives**

&#10102; Test Hypotheses about a Mean

&#10103; Explain the Difference between Statistical Significance and Practical Significance

---

### Objective 1: Test Hypotheses about a Mean

<u>OBJECTIVE 1, PAGE 1</u>

*Answer the following after watching the video that explains the procedure for testing hypotheses about a mean.*

1) What are the three conditions that must be satisfied before testing a hypothesis regarding a population mean, $\mu$?

2) State the five steps for testing a hypothesis about a population mean, $\mu$.

Step 1

Step 2

<u>OBJECTIVE 1, PAGE 1 (CONTINUED)</u>
Step 3 (By Hand)                                    Step 3 (Using Technology)




Step 4




Step 5




<u>OBJECTIVE 1, PAGE 2</u>
3) What tool is used to determine if the sample is drawn from a population that is normally distributed?




4) What tool is used to determine if the sample contains outliers?

**Example 1**     *Testing a Hypothesis about a Population Mean: Large Sample*

The mean height of American males is 69.5 inches. The heights of the 44 male U.S. presidents (Washington through Trump) have a mean of 70.84 inches and a standard deviation of 2.73 inches. Treating the 44 presidents as a simple random sample, determine whether there is evidence to suggest that U.S. presidents are taller than the average American male. Use the $\alpha = 0.05$ level of significance. (Note: Grover Cleveland was elected to two nonconsecutive terms, so technically there have been 45 presidents of the United States.)

| **Example 2** | ***Testing a Hypothesis about a Population Mean: Small Sample*** |

The "fun size" of a Snickers bar is supposed to weigh 20 grams. Because the penalty for selling candy bars under their advertised weight is severe, the manufacturer calibrates the machine so that the mean weight is 20.1 grams. The quality control engineer at Mars, the Snickers manufacturer, is concerned about the calibration. He obtains a random sample of 11 candy bars, weighs them, and obtains the data in Table 1. Should the machine be shut down and calibrated? Because shutting down the plant is expensive, he decides to conduct the test at the $\alpha = 0.01$ level of significance.

**Table 1**

| | | |
|-------|-------|-------|
| 19.68 | 20.66 | 19.56 |
| 19.98 | 20.65 | 19.61 |
| 20.55 | 20.36 | 21.02 |
| 21.50 | 19.74 | |

Data from Michael Carlisle, student at Joliet Junior College

## Objective 2: Explain the Difference between Statistical Significance and Practical Significance

OBJECTIVE 2, PAGE 1

5) What does practical significance refer to?

OBJECTIVE 2, PAGE 2

| Example 3 | Statistical versus Practical Significance |

According to the American Community Survey, the mean travel time to work in Collin County, Texas, is 27.6 minutes. The Department of Transportation reprogrammed all the traffic lights in Collin County in an attempt to reduce travel time. To determine whether there is evidence that travel time has decreased as a result of the reprogramming, the Department of Transportation obtains a random sample of 2500 commuters, records their travel time to work, and finds a sample mean of 27.3 minutes with a standard deviation of 8.5 minutes. Does this result suggest that travel time has decreased at the $\alpha = 0.05$ level of significance?

OBJECTIVE 2, PAGE 3

Large sample sizes can lead to results that are statistically significant, whereas the difference between the statistic and parameter in the null hypothesis is not enough to be considered practically significant.

## Section 10.3A
## Hypothesis Tests on a Population Mean Using Simulation and the Bootstrap

**Objectives**

❶ Test Hypotheses about a Population Mean Using the Simulation Method

❷ Test Hypotheses about a Population Mean Using the Bootstrap

INTRODUCTION, PAGE 1

1) In tests regarding the population mean, the null hypothesis will be $H_0$: $\mu = \mu_0$. What are the three possible alternative hypotheses?

The statement in the null hypothesis is assumed to be true and we are looking for evidence in support of the statement in the alternative hypothesis. Put another way, we want to know if a sample mean could come from a population whose mean is $\mu_0$ and any difference between $\mu_0$ and the sample mean is due to random chance. Or, is the sample mean from a population whose mean is different from (two-tailed), less than (left-tailed), or greater than (right-tailed) $\mu_0$?

INTRODUCTION, PAGE 2

Means are computed from quantitative data, so coins and urns are not going to be useful in simulating outcomes to build a null model that could be used to conduct inference on the mean. There are two approaches that may be utilized in building the null model when testing claims about a population mean – the simulation method and the Bootstrap.

---

### Objective 1: Test Hypotheses about a Population Mean Using the Simulation Method

OBJECTIVE 1, PAGE 1

2) In order to test hypotheses about a population mean, what are the requirements to use the simulation method to build the null model?

3) Explain how the simulation method is used to approximate the $P$-value.

**Example 1**     *Testing a Hypothesis about a Population Mean Using Simulation*

Coors Field is home to the Colorado Rockies baseball team and is located in Denver, Colorado. Denver is approximately one mile above sea level where the air is thinner. Therefore, baseballs are thought to travel farther in this stadium. Does the evidence support this belief? In a random sample of 15 home runs hit in Coors Field, the mean distance the ball traveled was 417.3 feet. Does this represent evidence to suggest that the ball travels farther in Coors Field than it does in the other Major League ballparks?

| | | | | |
|---|---|---|---|---|
| 427 | 383 | 399 | 444 | 414 |
| 397 | 421 | 395 | 427 | 432 |
| 399 | 415 | 433 | 427 | 446 |

<u>OBJECTIVE 1, PAGE 3</u>
4) State the four steps for testing a hypothesis about a population mean using simulation.


Step 1



Step 2



Step 3




Step 4










<u>OBJECTIVE 1, PAGE 4</u>
Note that the $P$-value for the hypothesis test $H_0 : \mu = 400.0$ feet versus $H_1 : \mu > 400.0$ feet using Student's $t$-distribution is 0.0017, which is very close to the approximate $P$-value using simulation of 0.002 we obtained in Example 1.

## Objective 2: Test Hypotheses about a Population Mean Using the Bootstrap

5) When using the bootstrap method, is the sampling done with replacement or without replacement?

6) In doing any hypothesis testing, we always generate the sampling distribution under the assumption the null hypothesis is true. When using bootstrapping to test hypotheses about a population mean, what first must be done to the sample data?

---

**Example 2**  *Using the Bootstrap Method To Test a Hypothesis about a Population Mean*

The "fun size" of a Snickers bar is supposed to weigh 20 grams. Because the penalty for selling candy bars under their advertised weight is severe, the manufacturer calibrates the machine so that the mean weight is 20.1 grams. The quality control engineer at Mars, the Snickers manufacturer, is concerned about the calibration. He obtains a random sample of 11 candy bars, weighs them, and obtains the data in Table 1. Should the machine be shut down and calibrated?

**Table 1**

| | | | |
|---|---|---|---|
| 19.68 | 20.66 | 19.56 | 21.50 |
| 19.98 | 20.65 | 19.61 | 19.74 |
| 20.55 | 20.36 | 21.02 | |

Data from Michael Carlisle, student at Joliet Junior College

---

<u>OBJECTIVE 2, PAGE 3</u>

7) State the four steps for testing a hypothesis about a population mean using simulation.

Step 1

Step 2

Step 3

Step 4

<u>OBJECTIVE 2, PAGE 4</u>

Note that the *P*-value for the hypothesis in Example 2 using Student's *t*-distribution is 0.3226. This is fairly close to the approximate *P*-value using the Bootstrap method.

## Section 10.4
## Putting It Together: Which Procedure Do I Use?

**Objective**

❶ Determine the Appropriate Hypothesis Test to Perform

---

### *Objective 1: Determine the Appropriate Hypothesis Test to Perform*

OBJECTIVE 1, PAGE 1

*Answer the following after watching the video.*

1) What is the type of the variable of interest when testing a population proportion, $p$?

2) List two of the conditions that must be met when testing a population proportion, $p$.

3) What is the type of the variable of interest when testing a population mean, $\mu$?

4) Besides the facts that the sample must be obtained by simple random sampling or through a randomized experiment and that the sample size must be small relative to the size of the population, what other condition must be satisfied?

<u>OBJECTIVE 1, PAGE 2</u>
**Flowchart for Determining Which Type of Test to Perform**

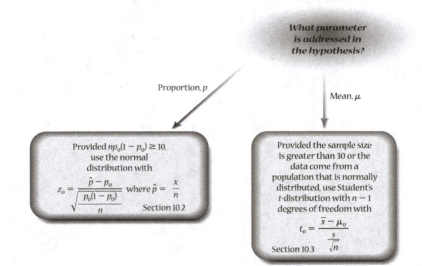

# Chapter 11 – Inference on Two Samples

## Putting It Together

In Chapters 9 and 10, we discussed inferences regarding a single population parameter. The inferential methods presented in those chapters will be modified slightly in this chapter so that we can compare two population parameters.

Section 11.1 presents inferential methods for comparing two population proportions. That is, inference when the response variable is qualitative with two possible outcomes (success or failure). The first order of business is to decide whether the data are obtained from an independent or dependent sample—simply put, we determine if the observations in one sample are somehow related to the observations in the other. We then discuss methods for comparing two proportions from independent samples. Methods for comparing two proportions from dependent samples are covered in Section 12.3.

Section 11.2 presents inferential methods used to handle dependent samples when the response variable is quantitative. For example, we might want to know whether the reaction time in an individual's dominant hand is different from the reaction time in the non-dominant hand.

Section 11.3 presents inferential methods used to handle independent samples when there are two levels of treatment and the response variable is quantitative. For example, we might randomly divide 100 volunteers who have a common cold into two groups. The control group would receive a placebo and the experimental group would receive an experimental drug. The response variable might be time until the cold symptoms go away.

We wrap up the chapter with a Putting It Together section. One of the more difficult aspects of inference is determining which inferential method to use. This section helps develop this skill.

## Section 11.1
## Inference about Two Population Proportions: Independent Samples

**Objectives**

❶ Distinguish between Independent and Dependent Sampling

❷ Test Hypotheses Regarding Two Population Proportions from Independent Samples

❸ Construct and Interpret Confidence Intervals for the Difference between Two Population Proportions

❹ Determine the Sample Size Necessary for Estimating the Difference between Two Population Proportions

---

### Objective 1: Distinguish between Independent and Dependent Sampling

OBJECTIVE 1, PAGE 1

*Answer the following after watching the video.*

1) Explain why the scenario involving the acne medications is an example of dependent sampling.

2) Explain why the scenario involving fast-food receipts is an example of dependent sampling.

3) What does it mean to say that a sampling method is independent?

4) What does it mean to say that a sampling method is dependent?

OBJECTIVE 1, PAGE 3

---

**Example 1**     *Distinguishing between Independent and Dependent Sampling*

Decide whether the sampling method is independent or dependent. Then determine whether the response variable is qualitative or quantitative.

A) Joliet Junior College decided to implement a course redesign of its developmental math program. Students enrolled in either a traditional lecture format course or a lab-based format course in which lectures and homework were done using video and the course management system MyMathLab. There were 1100 students enrolled in the traditional lecture format and 500 enrolled in the lab-based format. Once the course ended, the researchers determined whether the student passed the course. The goal of the study was to determine whether the proportion of students who passed the lab-based format exceeded that of students who passed the lecture format.

B) Do women tend to select a spouse who has an IQ higher than their own? To answer that question, researchers randomly selected 20 women and their husbands. They measured the IQ of each wife–husband team to determine whether there was a significant difference in IQ.

---

## *Objective 2: Test Hypotheses Regarding Two Population Proportions from Independent Samples*

<u>OBJECTIVE 2, PAGE 1</u>

**Sampling Distribution of the Difference between Two Proportions (Independent Sample)**

Suppose a simple random sample of size $n_1$ is taken from a population where $x_1$ of the individuals have a specified characteristic, and a simple random sample of size $n_2$ is independently taken from a different population where $x_2$ of the individuals have a specified characteristic.

The sampling distribution of $\hat{p}_1 - \hat{p}_2$, where $\hat{p}_1 = \dfrac{x_1}{n_1}$ and $\hat{p}_2 = \dfrac{x_2}{n_2}$, is approximately normal, with mean

$\mu_{\hat{p}_1 - \hat{p}_2} = p_1 - p_2$ and standard deviation $\sigma_{\hat{p}_1 - \hat{p}_2} = \sqrt{\dfrac{p_1(1-p_1)}{n_1} + \dfrac{p_2(1-p_2)}{n_2}}$, provided that

$n_1 \hat{p}_1 (1 - \hat{p}_1) \geq 10$ and $n_2 \hat{p}_2 (1 - \hat{p}_2) \geq 10$ and each sample size is no more than 5% of the population size.

The standardized version of $\hat{p}_1 - \hat{p}_2$ is then written as

$$Z = \frac{(\hat{p}_1 - \hat{p}_2) - (p_1 - p_2)}{\sqrt{\dfrac{p_1(1-p_1)}{n_1} + \dfrac{p_2(1-p_2)}{n_2}}},$$

which has an approximate standard normal distribution.

<u>OBJECTIVE 2, PAGE 2</u>

5) What are the three conditions that are necessary to test the difference between two population proportions using independent samples?

6) State the five steps for testing a hypothesis regarding the difference between two population proportions using independent samples.

Step 1

Step 2

Step 3 (By Hand)                                Step 3 (Using Technology)

Step 4

Step 5

**Example 2**     *Testing a Hypothesis Regarding Two Population Proportions*

In clinical trials of Nasonex®, 3774 adult and adolescent allergy patients (patients 12 years and older) were randomly divided into two groups. The patients in group 1 (experimental group) received 200µg of Nasonex, while the patients in group 2 (control group) received a placebo. Of the 2103 patients in the experimental group, 547 reported headaches as a side effect. Of the 1671 patients in the control group, 368 reported headaches as a side effect. It is known that over 10 million Americans who are 12 years and older are allergy sufferers. Is there significant evidence to conclude that the proportion of Nasonex users who experienced headaches as a side effect is greater than the proportion in the control group at the $\alpha = 0.05$ level of significance?

Looking back at the results of Example 2, we notice that the proportion of individuals taking 200 µg of Nasonex who experience headaches is statistically significantly greater than the proportion of individuals 12 years and older taking a placebo who experience headaches. While the difference of 4% is statistically significant, it does not have any practical significance.

***Objective 3: Construct and Interpret Confidence Intervals for the Difference between Two Population Proportions***

<u>OBJECTIVE 3, PAGE 1</u>

7) State the three conditions that are necessary to construct a $(1-\alpha)\cdot 100\%$ confidence interval for the difference between two population proportions from independent samples.

8) State the formulas for the lower bound and upper bound associated with a $(1-\alpha)\cdot 100\%$ confidence interval for the difference between two population proportions from independent samples.

| Example 3 | *Constructing a Confidence Interval for the Difference between Two Population Proportions* |

The Gallup organization surveyed 1100 adult Americans on May 6–9, 2002, and conducted an independent survey of 1100 adult Americans on May 3–7, 2017. In both surveys Gallup asked the following: "Right now, do you think the state of moral values in the country as a whole is getting better or getting worse?" On May 3–7, 2017, 846 of the 1100 surveyed responded that the state of moral values is getting worse; on May 6–9, 2002, 737 of the 1100 surveyed responded that the state of moral values is getting worse. Construct and interpret a 90% confidence interval for the difference between the two population proportions, $p_{2017} - p_{2002}$.

## Objective 4: Determine the Sample Size Necessary for Estimating the Difference between Two Population Proportions

<u>OBJECTIVE 4, PAGE 1</u>

9) State the formula for the margin of error, $E$, in constructing a $(1-\alpha)\cdot 100\%$ confidence interval for the difference between two population proportions.

<u>OBJECTIVE 4, PAGE 2</u>

10) State the formula for the sample size required to obtain a $(1-\alpha)\cdot 100\%$ confidence interval for the difference between two population proportions with a margin of error, $E$, if prior estimates of $p_1$ and $p_2$, $\hat{p}_1$ and $\hat{p}_2$, are available.

11) State the formula for the sample size required to obtain a $(1-\alpha)\cdot 100\%$ confidence interval for the difference between two population proportions with a margin of error, $E$, if prior estimates of $p_1$ and $p_2$ are not available.

Chapter 11: Inference on Two Samples

**Example 4**    *Determining Sample Size*

A nutritionist wants to estimate the difference between the proportion of males and females who consume the USDA's recommended daily intake of calcium.

A) What sample size should be obtained if she wants the estimate to be within 3 percentage points with 95% confidence, assuming that she uses the results of the USDA's 1994–1996 Diet and Health Knowledge Survey, according to which 51.1% of males and 75.2% of females consume the USDA's recommended daily intake of calcium.

B) What sample size should be obtained if she wants the estimate to be within 3 percentage points with 95% confidence, assuming that she does not use any prior estimates?

## Section 11.1A
## Using Randomization Techniques to Compare Two Proportions

**Objectives**

❶ Use Randomization to Compare Two Population Proportions

INTRODUCTION, PAGE 1

Recall in a completely randomized design, a group of individuals is randomly assigned to two or more treatment groups, the treatment is imposed on the individuals, and a response variable is measured. The methods of this section apply to completely randomized designs where there are two levels of the treatment (or two distinct groups) and the response variable is qualitative with two possible outcomes. Because the response variable is qualitative with two possible outcomes, we analyze the data using proportions.

*Objective 1: Use Randomization to Compare Two Population Proportions*

OBJECTIVE 1, PAGE 1

*Watch the video to gain an understanding of how randomization is used to compare two population proportions.*

1) What is the response variable in the study? Is it qualitative or quantitative?

2) What is the assumption about the group of students in the two courses?

OBJECTIVE 1, PAGE 2

3) List the two sample proportions. What is the difference between the two sample proportions?

Copyright © 2019 Pearson Education, Inc.

OBJECTIVE 1, PAGE 3

4) What are the two possibilities regarding the difference between the sample proportions?

5) State the two possibilities using the notation of hypothesis tests.

OBJECTIVE 1, PAGE 4

6) What is the difference in pass rate after randomly assigning students to a treatment? Does this suggest that students did better in the MRP course or the traditional course?

OBJECTIVE 1, PAGE 5

7) What is the difference in pass rate for each of the two random assignments using StatCrunch's Urn applet? Did either result in a difference of pass rates as extreme, or more extreme, than the observed result of 0.209?

OBJECTIVE 1, PAGE 6

8) After 2000 more random assignments, what proportion of the times did we observe with 19 or more students passing the MRP course? Based on these results, what is the appropriate decision about the null hypothesis?

OBJECTIVE 1, PAGE 9

Rather than using the Urn applet to do the random assignment, we can use the Randomization Test for Two Proportions applet in StatCrunch.

OBJECTIVE 1, PAGE 10

9) Examine the graph of the outcomes in the random assignments of students. Where is the graph centered? Why should this not be surprising?

10) What is the shape of the distribution of the difference in pass rates?

OBJECTIVE 1, PAGE 11

11) Suppose we wanted to determine if there was a difference in the pass rates of the MRP course versus the Traditional course. What would the null and alternative hypotheses be?

OBJECTIVE 1, PAGE 12
12) State the five steps for testing hypotheses regarding two proportions using random assignment.

Step 1

Step 2

Step 3

Step 4

Step 5

**Example 2**   *Testing a Hypothesis Regarding Two Population Proportions*

Zoloft and Trintellix are drugs meant to alleviate symptoms associated with major depressive disorder. In clinical trials of 3066 patients taking Zoloft, 790 reported nausea as a side effect. In clinical trials of 1013 patients taking Trintellix, 216 reported nausea as a side effect. Is there significant evidence to conclude a difference in the proportion of patients who report nausea as a side effect for these two drugs?
Source: www.zoloft.com and us.trintellix.com.

## Section 11.2
## Inference about Two Population Means: Dependent Samples

### Objectives

❶ Test Hypotheses for a Population Mean from Matched-Pairs Data

❷ Construct and Interpret Confidence Intervals about a Population Mean Difference of Matched-Pairs Data

---

*Objective 1: Test Hypotheses for a Population Mean from Matched-Pairs Data*

OBJECTIVE 1, PAGE 1

Inference on matched-pairs data is similar to inference regarding a single population mean.

Recall that if the population from which the sample was drawn is normally distributed with no outliers or the sample size is large $(n \geq 30)$, we said that

$$t = \frac{\overline{x} - \mu}{\frac{s}{\sqrt{n}}}$$

follows Student's $t$-distribution with $n-1$ degrees of freedom.

OBJECTIVE 1, PAGE 2

*Watch the video that explains the procedure for hypothesis tests regarding the difference between two dependent means.*

1) What are the four conditions that must be satisfied before testing a hypothesis regarding the difference between two population means using dependent samples?

OBJECTIVE 1, PAGE 2 (CONTINUED)
2) State the five steps for testing a hypothesis regarding the difference between two population means using dependent samples.

Step 1

Step 2

Step 3 (By Hand)                                          Step 3 (Using Technology)

Step 4

Step 5

OBJECTIVE 1, PAGE 3
3) What tool is used to determine if the differenced data come from a population that is normally distributed?

4) What tool is used to determine if the differenced data contain outliers?

**Example 1**     *Testing a Hypothesis for Matched-Pairs Data*

Professor Andy Neill measured the time (in seconds) required to catch a falling meter stick for 12 randomly selected students' dominant hand and non-dominant hand. Professor Neill wants to know if the reaction time in an individual's dominant hand is less than the reaction time in his or her non-dominant hand. A coin flip is used to determine whether reaction time is measured using the dominant or non-dominant hand first. Conduct the test at the $\alpha = 0.05$ level of significance. The data obtained are presented in Table 1.  (Data from Professor Andy Neill, Joliet Junior College)

| **Table 1** | Student | Dominant Hand, $X_i$ | Non-dominant Hand, $Y_i$ |
|---|---|---|---|
| | 1 | 0.177 | 0.179 |
| | 2 | 0.210 | 0.202 |
| | 3 | 0.186 | 0.208 |
| | 4 | 0.189 | 0.184 |
| | 5 | 0.198 | 0.215 |
| | 6 | 0.194 | 0.193 |
| | 7 | 0.160 | 0.194 |
| | 8 | 0.163 | 0.160 |
| | 9 | 0.166 | 0.209 |
| | 10 | 0.152 | 0.164 |
| | 11 | 0.190 | 0.210 |
| | 12 | 0.172 | 0.197 |

## Objective 2: Construct and Interpret Confidence Intervals about a Population Mean Difference of Matched-Pairs Data

OBJECTIVE 2, PAGE 1

5) State the formulas for the lower bound and upper bound associated with a $(1-\alpha)\cdot 100\%$ confidence interval for the population mean difference $\mu_d$.

OBJECTIVE 2, PAGE 2

**Example 2**     *Constructing a Confidence Interval for Matched-Pairs Data*

Using the data from Table 1, construct a 95% confidence interval estimate of the mean difference, $\mu_d$.

**Table 1**

| Student | Dominant Hand, $X_i$ | Non-dominant Hand, $Y_i$ |
|---|---|---|
| 1 | 0.177 | 0.179 |
| 2 | 0.210 | 0.202 |
| 3 | 0.186 | 0.208 |
| 4 | 0.189 | 0.184 |
| 5 | 0.198 | 0.215 |
| 6 | 0.194 | 0.193 |
| 7 | 0.160 | 0.194 |
| 8 | 0.163 | 0.160 |
| 9 | 0.166 | 0.209 |
| 10 | 0.152 | 0.164 |
| 11 | 0.190 | 0.210 |
| 12 | 0.172 | 0.197 |

Data from Professor Andy Neill, Joliet Junior College

## Section 11.2A
## Using Bootstrapping to Conduct Inference on Two Dependent Means

**Objectives**

❶ Test Hypotheses about Two Dependent Means Using the Bootstrap Method

INTRODUCTION, PAGE 1

1) How are the pairs selected in a matched-pair experimental design?

2) In the matched-pairs design, how many levels of treatment are there?

*Objective 1: Test Hypotheses about Two Dependent Means Using the Bootstrap Method*

OBJECTIVE 1, PAGE 1

Earlier, we presented the Bootstrap method for testing hypotheses about a single mean. We can use this method to test hypotheses on matched-pairs data. To use the Bootstrap, compute the difference in each matched pair. As with Bootstrapping on a single mean, the differenced data must first be adjusted to build the null model so that the data come from a population whose mean difference is that stated in the null hypothesis. Then find bootstrap resamples based on the differenced data.

3) List the three ways the null and alternative hypotheses can be structured when analyzing matcher pairs data.

**Example 1**     *Using Bootstrapping to Test Hypotheses about Matched-Pairs Data*

Professor Andy Neill measured the time (in seconds) required to catch a falling meter stick for 12 randomly selected students' dominant hand and non-dominant hand. Professor Neill wants to know if the reaction time in an individual's dominant hand is less than the reaction time in his or her non-dominant hand. A coin flip is used to determine whether reaction time is measured using the dominant or non-dominant hand first. The data obtained are presented in Table 1.

| **Table 1** | Student | Dominant Hand, $X_i$ | Non-dominant Hand, $Y_i$ |
|---|---|---|---|
| | 1 | 0.177 | 0.179 |
| | 2 | 0.210 | 0.202 |
| | 3 | 0.186 | 0.208 |
| | 4 | 0.189 | 0.184 |
| | 5 | 0.198 | 0.215 |
| | 6 | 0.194 | 0.193 |
| | 7 | 0.160 | 0.194 |
| | 8 | 0.163 | 0.160 |
| | 9 | 0.166 | 0.209 |
| | 10 | 0.152 | 0.164 |
| | 11 | 0.190 | 0.210 |
| | 12 | 0.172 | 0.197 |

Data from Professor Andy Neill, Joliet Junior College

<u>OBJECTIVE 1, PAGE 3</u>
4) List the six steps for using the bootstrap to test hypotheses about matched-pairs data.

Step 1

Step 2

Step 3

Step 4

Step 5

Step 6

## Section 11.3
## Inference about Two Population Means: Independent Samples

**Objectives**

❶ Test Hypotheses Regarding Two Population Means from Independent Samples

❷ Construct and Interpret Confidence Intervals about the Difference of Two Independent Means

---

*Objective 1: Test Hypotheses Regarding Two Population Means from Independent Samples*

OBJECTIVE 1, PAGE 1

1) For the study about the new experimental drug, what is the response variable? Is it qualitative or quantitative?

2) If we let $\mu_1$ represent the mean time until cold symptoms go away for the individuals taking the drug and $\mu_2$ represent the mean time until cold symptoms go away, what are the null and alternative hypotheses?

OBJECTIVE 1, PAGE 2

**Sampling Distribution of the Difference of Two Means: Independent Samples with Population Standard Deviations Unknown (Welch's *t*)**

Suppose that a simple random sample of size $n_1$ is taken from a population with unknown mean $\mu_1$ and unknown standard deviation $\sigma_1$. In addition, a simple random sample of size $n_2$ is taken from a second population with unknown mean $\mu_2$ and unknown standard deviation $\sigma_2$. If the two populations are normally distributed or the sample sizes are sufficiently large $(n_1 \geq 30 \text{ and } n_2 \geq 30)$, then

$$t_0 = \frac{(\bar{x}_1 - \bar{x}_2) - (\mu_1 - \mu_2)}{\sqrt{\dfrac{s_1^2}{n_1} + \dfrac{s_2^2}{n_2}}}$$

approximately follows Student's *t*-distribution with the smaller of $n_1 - 1$ or $n_2 - 1$ degrees of freedom, where $\bar{x}_1$ is the sample mean and $s_1$ is the sample standard deviation from population 1, and $\bar{x}_2$ is the sample mean and $s_2$ is the sample standard deviation from population 2.

OBJECTIVE 1, PAGE 3

3) What are the four conditions that must be satisfied before testing a hypothesis regarding the difference between two population means using independent samples?

4) State the five steps for testing a hypothesis regarding the difference between two population means using independent samples.

Step 1

Step 2

Step 3 (By Hand)                     Step 3 (Using Technology)

Step 4

Step 5

5) What tool is used to determine if the sample is drawn from a population that is normally distributed?

6) What tool is used to determine if the sample contains outliers?

**Example 1**     *Testing a Hypothesis Regarding Two Independent Means*

In the Spacelab Life Sciences 2 payload, 14 male rats were sent to space. Upon their return, the red blood cell mass (in milliliters) of the rats was determined. A control group of 14 male rats was held under the same conditions (except for space flight) as the space rats, and their red blood cell mass was also determined when the space rats returned. The project, led by Dr. Paul X. Callahan, resulted in the data listed in Table 2. Does the evidence suggest that the flight animals have a different red blood cell mass from that of the control animals at the $\alpha = 0.05$ level of significance?

**Table 2**

| Flight | | | | Control | | | |
|---|---|---|---|---|---|---|---|
| 8.59 | 8.64 | 7.43 | 7.21 | 8.65 | 6.99 | 8.40 | 9.66 |
| 6.87 | 7.89 | 9.79 | 6.85 | 7.62 | 7.44 | 8.55 | 8.70 |
| 7.00 | 8.80 | 9.30 | 8.03 | 7.33 | 8.58 | 9.88 | 9.94 |
| 6.39 | 7.54 | | | 7.14 | 9.14 | | |

Data from NASA Life Sciences Data Archive

<u>OBJECTIVE 1, PAGE 5 (CONTINUED)</u>

<u>OBJECTIVE 1, PAGE 6</u>

7) Using the smaller of $n_1 - 1$ or $n_2 - 1$ for the degrees of freedom is conservative. State the formula for degrees of freedom that is used by computer software for increased precision.

---

## Objective 2: Construct and Interpret Confidence Intervals about the Difference of Two Independent Means

<u>OBJECTIVE 2, PAGE 1</u>

8) State the formulas for the lower bound and upper bound associated with a $(1-\alpha) \cdot 100\%$ confidence interval for the difference of two means.

---

**Example 2**     *Constructing a Confidence Interval for the Difference of Two Independent Means*

Recently, a device called low-level laser therapy has evolved as a potential solution to hair loss. A randomized, placebo-controlled experiment was conducted in which 65 females were randomly assigned to one of two treatment groups. The 43 subjects in Group 1 were exposed to a 9-beam lasercomb treatment, while the 22 subjects in Group 2 received a placebo treatment for a total of 16 weeks. The response variable in the study was hair density (measured in hair count per square centimeter). For the subjects in Group 1, the mean change in hair density was 20.2 with a standard deviation of 11.2. For the subjects in Group 2, the mean change in hair density was 2.8 with a standard deviation of 16.5. Estimate the mean difference in hair density between Group 1 and Group 2 by constructing a 95% confidence interval. Note: Analysis of the sample data suggest the sample data come from populations that are normally distributed. Source: Jimenez, JJ, et. al. "Efficacy and safety of low-level laser device in the treatment of male and female pattern hair loss: a multicenter, randomized, sham device-controlled, double-blind study." American Journal of Clinical Dermatology Volume 15, Issue 2(2014).

---

Statistical software provides an option for two types of two-sample *t*-tests: one that assumes equal population variances (pooling) and another that does not assume equal population variances. Welch's *t*-statistic does not assume that the population variances are equal.

Because testing the equality of variances is so volatile, we are content to use Welch's *t*. Welch's *t*-test is more conservative than the pooled *t*. The price that must be paid for the conservative approach is that the probability of a Type II error is higher with Welch's *t* than with the pooled *t* when the population variances are equal. However, the two tests typically provide the same conclusion.

## Section 11.3A
## Using Randomization Techniques to Compare Two Independent Means

**Objectives**

❶ Use Randomization to Compare Two Population Means

---

### Objective 1: Use Randomization to Compare Two Population Means

<u>OBJECTIVE 1, PAGE 1</u>

1) For the study, what is the response variable? Is it qualitative or quantitative?

<u>OBJECTIVE 1, PAGE 3</u>

2) View the dot plot by gender. Does it suggest that females are spending more time on homework, on average?

<u>OBJECTIVE 1, PAGE 3</u>

3) What are the two possible explanations for the sample mean difference of 19.2 minutes?

<u>OBJECTIVE 1, PAGE 4</u>

4) If $\mu_1$ is the mean time spent on homework by female students and $\mu_2$ is the mean amount of time spent on homework by male students, what are the null and alternative hypotheses?

OBJECTIVE 1, PAGE 5
5) What is the sample mean difference (females minus males) for the set of randomly assigned data?

OBJECTIVE 1, PAGE 6
6) What is the sample mean difference (females minus males) for the set of randomly assigned data from StatCrunch?

OBJECTIVE 1, PAGE 7
7) Only 31 out of 5000 random assignments of gender to study time produced a sample mean difference of 19.2 minutes or higher. What is the *P*-value for this hypothesis test?

8) Based on the observed results, what is our decision about the statement in the null hypothesis? What is the conclusion for this test?

OBJECTIVE 1, PAGE 8
9) What is the shape of the distribution of randomized differences?

10) Where is the distribution of randomized differences centered? Why should this not be surprising?

11) State the five steps for testing hypotheses regarding two independent means using random assignment.
Step 1



Step 2




Step 3




Step 4




Step 5

**Example 1**    *Testing Hypotheses Regarding Two Independent Means*

In the Spacelab Life Sciences 2 payload, 14 male rats were sent to space. Upon their return, the red blood cell mass (in milliliters) of the rats was determined. A control group of 14 male rats was held under the same conditions (except for space flight) as the space rats, and their red blood cell mass was also determined when the space rats returned. The project, led by Dr. Paul X. Callahan, resulted in the data listed in Table 2. Does the evidence suggest that the flight animals have a different red blood cell mass from that of the control animals at the $\alpha = 0.05$ level of significance?

**Table 2**

| Flight | | | | Control | | | |
|------|------|------|------|------|------|------|------|
| 8.59 | 8.64 | 7.43 | 7.21 | 8.65 | 6.99 | 8.40 | 9.66 |
| 6.87 | 7.89 | 9.79 | 6.85 | 7.62 | 7.44 | 8.55 | 8.70 |
| 7.00 | 8.80 | 9.30 | 8.03 | 7.33 | 8.58 | 9.88 | 9.94 |
| 6.39 | 7.54 | | | 7.14 | 9.14 | | |

Data from NASA Life Sciences Data Archive

## Section 11.4
## Putting It Together: Which Procedure Do I Use?

**Objective**

❶ Determine the Appropriate Hypothesis Test to Perform

---

*Objective 1: Determine the Appropriate Hypothesis Test to Perform*

OBJECTIVE 1, PAGE 1

*Answer the following after watching the video.*

1) How can you determine whether a scenario calls for inference using proportions or inference using means?

2) Explain how to determine whether two samples are dependent or independent?

**Flowchart for Determining Which Type of Test to Perform**

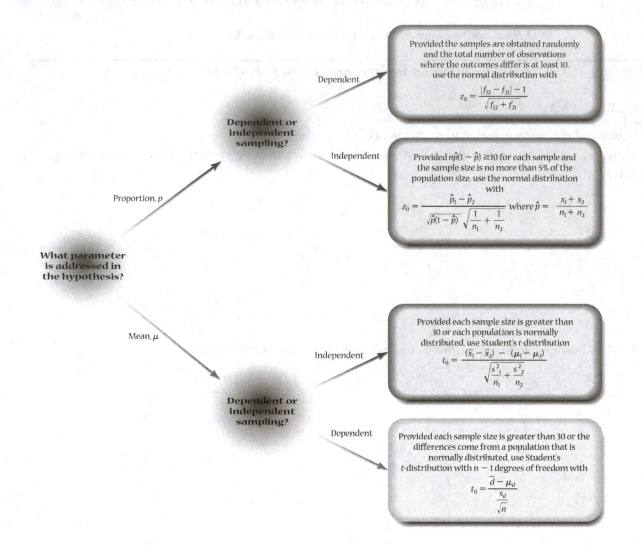

# Chapter 12 – Inference on Categorical Data

## OUTLINE

**Putting It Together**

In Chapters 9 through 11, we introduced statistical methods for testing hypotheses regarding a parameter such as $p$ or $\mu$.

Often, however, rather than testing a hypothesis regarding a parameter of a probability distribution, we want to test a hypothesis regarding the entire probability distribution. For example, we might test if the distribution of colors in a bag of plain M&M candies is 13% brown, 14% yellow, 13% red, 20% orange, 24% blue, and 16% green. Methods for testing such hypotheses are covered in Section 12.1.

In Section 12.2, we discuss a method for determining whether two qualitative variables are independent based on a sample. If they are not independent, the value of one variable affects the value of the other variable, and the variables are related. We conclude Section 12.2 by introducing tests for homogeneity, which compare proportions from two or more populations. This test are an extension of the two-sample $z$-test for proportions from independent samples discussed in Section 11.1.

We end the chapter with a discussion for comparing two proportions from dependent samples in Section 12.3.

## Section 12.1
## Goodness-of-Fit Test

**Objective**

❶ Perform a Goodness-of-Fit Test

INTRODUCTION, PAGE 2

1) State the characteristics of the chi-square distribution.

INTRODUCTION, PAGE 4

2) What does $\chi_\alpha^{\,2}$ represent?

INTRODUCTION, PAGE 5

| **Example 1** | ***Finding Critical Values for the Chi-Square Distribution*** |

Find the critical values that separate the middle 90% of the chi-square distribution from the 5% area in each tail, assuming 15 degrees of freedom.

3) Explain how to determine which row in Table VIII to use if the number of degrees of freedom is not in the table.

## Objective 1: Perform a Goodness-of-Fit Test

OBJECTIVE 1, PAGE 1
4) State the definition of a goodness-of-fit test.

OBJECTIVE 1, PAGE 2
5) Explain how to obtain the expected counts for a goodness-of-fit test.

| Example 2 | *Finding Expected Counts* |

One growing concern regarding the U.S. economy is the inequality in the distribution of income. The data in Table 1 represents the distribution of household income for various levels of income in 2000. An economist would like to know if income distribution is changing, so she randomly selects 1500 households and obtains the household income. Find the expected number of households in each income level assuming that the distribution of income has not changed since 2000.
**Note:** The income data has been adjusted for inflation.

**Table 1**
Distribution of income in the U.S. in 2000

| Income | Percent |
|---|---|
| Under $15,000 | 10.2 |
| $15,000 to $24,999 | 9.8 |
| $25,000 to $34,999 | 9.9 |
| $35,000 to $49,999 | 13.3 |
| $50,000 to $74,999 | 18.1 |
| $75,000 to $99,999 | 13.1 |
| $100,000 to $149,999 | 14.6 |
| $150,000 to $199,999 | 5.8 |
| At least $200,000 | 5.2 |

Data from U.S. Census Bureau

<u>Objective 1, Page 6</u>
6) State the test statistic for a goodness-of-fit test.

7) What two conditions are needed in order to use the test statistic $\chi^2 = \sum \frac{(O_i - E_i)^2}{E_i}$ ?

8) State the five steps for a goodness-of-fit test.

Step 1

Step 2

Step 3

Step 4

Step 5

---

**Example 3**    *Conducting a Goodness-of-Fit Test*

One growing concern regarding the U.S. economy is the inequality in the distribution of income. The data in Table 2 represents the distribution of household income in 2000. An economist would like to know if income distribution is changing, so she randomly selected 1500 households and obtained the household income. Table 3 contains the results of this survey

Note: The data in Table 3 are based on the 2016 Current Population Survey and have been adjusted for inflation.

**Table 2**

| Income | Percent |
|---|---|
| Under $15,000 | 10.2 |
| $15,000 to $24,999 | 9.8 |
| $25,000 to $34,999 | 9.9 |
| $35,000 to $49,999 | 13.3 |
| $50,000 to $74,999 | 18.1 |
| $75,000 to $99,999 | 13.1 |
| $100,000 to $149,999 | 14.6 |
| $150,000 to $199,999 | 5.8 |
| At least $200,000 | 5.2 |

**Table 3**

| Income | Frequency |
|---|---|
| Under $15,000 | 168 |
| $15,000 to $24,999 | 144 |
| $25,000 to $34,999 | 141 |
| $35,000 to $49,999 | 194 |
| $50,000 to $74,999 | 254 |
| $75,000 to $99,999 | 184 |
| $100,000 to $149,999 | 210 |
| $150,000 to $199,999 | 99 |
| At least $200,000 | 106 |

Does the evidence suggest that the distribution of income has changed since 2000 at the $\alpha = 0.05$ level of significance?

Chapter 12: Inference on Categorical Data

OBJECTIVE 1, PAGE 9
In Example 3, if we compare the observed and expected counts, we see where the shift in income is occurring.

OBJECTIVE 1, PAGE 11

**Example 4** *Conducting a Goodness-of-Fit Test*
An obstetrician wants to know whether the proportion of children born on each day of the week are the same. She randomly selects 500 birth records and obtains the data shown in Table 4 (Data from Vital Statistics of the United States).
Is there reason to believe that the day on which a child is born occurs with equal frequency at the $\alpha = 0.01$ level of significance?

**Table 4**

| Day of Week | Frequency |
| --- | --- |
| Sunday | 46 |
| Monday | 76 |
| Tuesday | 83 |
| Wednesday | 81 |
| Thursday | 81 |
| Friday | 80 |
| Saturday | 53 |

OBJECTIVE 1, PAGE 13
Goodness-of-Fit tests cannot be used to test whether sample data follow a specific distribution. We can only say the data is consistent with a distribution stated in the null hypothesi

## Section 12.2
## Tests for Independence and the Homogeneity of Proportions

**Objectives**

❶ Perform a Test for Independence

❷ Perform a Test for Homogeneity of Proportions

*Objective 1: Perform a Test for Independence*

<u>OBJECTIVE 1, PAGE 1</u>

1) What do the data presented in a contingency table measure?

<u>OBJECTIVE 1, PAGE 2</u>

2) What is the chi-square test for independence used to determine? Describe the null and alternative hypotheses for a chi-square test for independence.

OBJECTIVE 1, PAGE 3

| Example 1 | *Determining the Expected Counts in a Test for Independence* |
|---|---|

Is there a relationship between marital status and happiness? The data in Table 5 show the marital status and happiness of individuals who participated in the General Social Survey. Compute the expected counts within each cell, assuming that marital status and happiness are independent.

**Table 5**

|  |  | Marital Status | | | |
|---|---|---|---|---|---|
|  |  | Married | Widowed | Divorced/ Separated | Never Married |
| Happiness | Very Happy | 600 | 63 | 112 | 144 |
|  | Pretty Happy | 720 | 142 | 355 | 459 |
|  | Not Too Happy | 93 | 51 | 119 | 127 |

OBJECTIVE 1, PAGE 4
3) State the formula for finding expected frequencies in a chi-square test for independence.

OBJECTIVE 1, PAGE 6
4) State the test statistic for a chi-square test for independence.

OBJECTIVE 1, PAGE 6 (CONTINUED)

5) What two conditions are needed in order to use the test statistic $\chi^2 = \sum \dfrac{(O_i - E_i)^2}{E_i}$?

OBJECTIVE 1, PAGE 7

6) State the five steps for a test for independence.

Step 1

Step 2

Step 3

Step 4

Step 5

**Example 2**    *Performing a Chi-Square Test for Independence*

Does one's happiness depend on one's marital status? The data in Table 5 represent the marital status and disclosed level of happiness for a random sample of adult Americans who participated in the General Social Survey. Use this data to answer the question at the $\alpha = 0.05$ level of significance.

**Table 5**

|  |  | Married | Widowed | Divorced/ Separated | Never Married |
|---|---|---|---|---|---|
| Happiness | Very Happy | 600 | 63 | 112 | 144 |
|  | Pretty Happy | 720 | 142 | 355 | 459 |
|  | Not Too Happy | 93 | 51 | 119 | 127 |

The "Marital Status" label spans the four columns: Married, Widowed, Divorced/Separated, Never Married.

OBJECTIVE 1, PAGE 11
**Visual Support of Conclusions of Inference**
To see the association between two qualitative variables, draw bar graphs of the conditional distributions of the response variable by the explanatory variable.

OBJECTIVE 1, PAGE 12

**Example 3**    *Constructing a Conditional Distribution and Bar Graph*

Find the conditional distribution of happiness by marital status for the data in Table 5. Then draw a bar graph that represents the conditional distribution of happiness by marital status.

**Table 5**

| | | Marital Status | | | |
|---|---|---|---|---|---|
| | | Married | Widowed | Divorced/ Separated | Never Married |
| Happiness | Very Happy | 600 | 63 | 112 | 144 |
| | Pretty Happy | 720 | 142 | 355 | 459 |
| | Not Too Happy | 93 | 51 | 119 | 127 |

## *Objective 2: Perform a Test for Homogeneity of Proportions*

<u>OBJECTIVE 2, PAGE 1</u>

7) Explain what the chi-square test for homogeneity tests.

<u>OBJECTIVE 2, PAGE 2</u>

8) The procedures for the test for independence and the test of homogeneity of proportions are the same. Explain how the data differ for the test for independence and the test for homogeneity differ.

---

**Example 4**      *A Test for Homogeneity of Proportions*

Zocor is a drug (Merck and Co.) that is meant to reduce the level of LDL (bad) cholesterol and increase the level of HDL (good) cholesterol. In clinical trials of the drug, patients were randomly divided into three groups. Group 1 received Zocor, group 2 received a placebo, and group 3 received cholestyramine, a cholesterol-lowering drug currently available. Table 7 contains the number of patients in each group who did and did not experience abdominal pain as a side effect. Is there evidence to indicate that the proportion of subjects in each group who experienced abdominal pain is different at the $\alpha = 0.01$ level of significance?

| Table 7 | Group 1 (Zocor) | Group 2 (placebo) | Group 3 (cholestyramine) |
|---|---|---|---|
| Number of people who experienced abdominal pain | 51 | 5 | 16 |
| Number of people who did not experience abdominal pain | 1532 | 152 | 163 |

Data from Merck and Co.

---

9) In the requirements for performing a chi-square test are not satisfied, list two options that a researcher has.

## Section 12.3
## Inference about Two Population Proportions: Dependent Samples

**Objective**

❶ Test Hypotheses Regarding Two Proportions from Dependent Samples

---

INTRODUCTION, PAGE 1

1) What is the name of the test used to compare two proportions with matcher-pairs data?

---

*Objective 1: Test Hypotheses Regarding Two Proportions from Dependent Samples*

OBJECTIVE 1, PAGE 1

McNemar's Test will be used to determine whether the difference in sample proportions is due to sampling error or whether there the differences are significant enough to conclude that the proportions are different. The null hypothesis is $H_0 : p_1 = p_2$, and the alternative hypothesis is $H_1 : p_1 \neq p_2$.

OBJECTIVE 1, PAGE 2

2) To use McNemar's test, explain how to arrange the data in a contingency table.

3) What two conditions are necessary for performing McNemar's test?

OBJECTIVE 1, PAGE 2 (CONTINUED)
4) State the five steps for McNemar's test.

Step 1

Step 2

Step 3 (By Hand)

Step 3 (Using Technology)

Step 4

Step 5

**Example 1**     *Analyzing the Difference of Two Proportions: Dependent Samples*

A recent General Social Survey asked the following two questions of a random sample of 1492 adult Americans under the hypothetical scenario that the government suspected that a terrorist act was about to happen:

   Do you believe the authorities should have the right to tap people's telephone conversations?
   Do you believe the authorities should have the right to stop and search people on the street at random?

The results of the survey are shown in Table 9.

**Table 9**

|  |  | Random Stop | |
|---|---|---|---|
|  |  | Agree | Disagree |
| Tap Phone | Agree | 494 | 335 |
|  | Disagree | 126 | 537 |

Do the proportions who agree with each scenario differ significantly? Use the $\alpha=0.05$ level of significance.

# Chapter 13 – Comparing Three or More Means

## OUTLINE

**Putting It Together**

Do you remember the progression for comparing proportions? Chapters 9 and 10 discussed inference for a single proportion, Chapter 11 discussed inference for two proportions, and Chapter 12 presented a discussion of inference for three or more proportions (homogeneity of proportions).

We have this same progression of topics for inference on means. In Chapters 9 and 10, we discussed inferential techniques for a single population mean. In Chapter 11, we discussed inferential techniques for comparing two means. In this chapter, we learn inferential techniques for comparing three or more means.

Just as we used a different distribution to compare multiple proportions (the chi-square distribution), we use a different distribution for comparing three or more means. Although the choice of distribution initially may seem strange, once the logic of the procedure is understood, the choice of distribution makes sense.

## Section 13.1
## Comparing Three or More Means: One-Way Analysis of Variance

**Objectives**

&#10112; Verify the Requirements to Perform a One-Way ANOVA

&#10113; Test a Hypothesis Regarding Three or More Means Using One-Way ANOVA

---

INTRODUCTION, PAGE 2

1) What is Analysis of Variance (ANOVA) used to test?

2) What are the null and alternative hypotheses for testing a hypothesis regarding three population means?

INTRODUCTION, PAGE 3

3) Sketch an example of what the distribution of the populations might look like if the alternative hypothesis (at least one population mean is different from the others) is true.

INTRODUCTION, PAGE 4

4) Why do we use ANOVA to test the equality of three or more means rather than comparing them two at a time using the techniques used in Section 11.3??

**Notes on Analysis of Variance**

Sir Ronald A. Fisher (1890–1962) introduced the method called analysis of variance (ANOVA). This term may seem odd because we are conducting a test on means, not variances. However, the name refers to the approach we are using, which will involve a comparison of two estimates of the same population variance. The justification for the name will become clear as we develop the test statistic.

The procedure used in this section is called one-way analysis of variance because only one factor distinguishes the various populations in the study.

We use a one-way analysis of variance when analyzing data from a completely randomized design with three or more levels of treatment. In this design, the subjects must be similar in all characteristics except the level of the treatment.

---

*Objective 1: Verify the Requirements to Perform a One-Way ANOVA*

OBJECTIVE 1, PAGE 1

5) State the four requirements to perform a one-way ANOVA test.

OBJECTIVE 1, PAGE 2

6) Explain how to verify the requirement of normality for a one-way ANOVA test.

7) State the rule of thumb for verifying the requirement of equal population variances for a one-way ANOVA test.

**Example 1**    *Verifying the Requirements of One-Way ANOVA*

Prosthodentists specialize in the restoration of oral function, including the use of dental implants, veneers, dentures, and crowns. Because repair of chipped veneer is less costly and time-consuming than complete restoration, a researcher wanted to compare the shear bond strength of different kits for repairs of chipped porcelain veneer in fixed prosthodontics. He randomly divided 20 porcelain specimens into four treatment groups. Group 1 specimens used the Cojet system, group 2 used the Silistor system, group 3 used the Cimara system, and group 4 specimens used the Ceramic Repair system. At the conclusion of the study, shear bond strength (in megapascals, MPa) was measured according to ISO 10477. The data in Table 1 are based on the results of the study. Verify that the requirements to perform one-way ANOVA are satisfied.

**Table 1**

| Cojet | Silistor | Cimara | Ceramic Repair |
|-------|----------|--------|----------------|
| 15.4  | 17.2     | 5.5    | 11.0           |
| 12.9  | 14.3     | 7.7    | 12.4           |
| 17.2  | 17.6     | 12.2   | 13.5           |
| 16.6  | 21.6     | 11.4   | 8.9            |
| 19.3  | 17.5     | 16.4   | 8.1            |

Data from P. Schmage et al. "Shear Bond Strengths of Five Intraoral Porcelain Repair Systems," Journal of Adhesion Science & Technology 21 (5–6):409–422, 2007

---

*Objective 2: Test a Hypothesis Regarding Three or More Means Using One-Way ANOVA*

*Answer the following after watching the video.*

8) Explain what is meant by the terms between-sample variability and within-sample variability.

9) What evidence will suggest that the samples come from populations with different means?

OBJECTIVE 2, PAGE 3

*Watch the video to learn how to compute the F-test statistic.*

OBJECTIVE 2, PAGE 4

10) State the formulas for the mean square due to error (MSE), the mean square due to treatment (MST), and the test statistic for a one-way ANOVA test $(F_0)$.

OBJECTIVE 2, PAGE 5

11) List the six steps for computing the F-test statistic by hand.

**Example 2**        *Computing the F-Test Statistic by Hand*

Compute the F-test statistic for the data shown

| $x_1$ | $x_2$ | $x_3$ |
|-------|-------|-------|
| 4     | 7     | 10    |
| 5     | 8     | 10    |
| 6     | 9     | 11    |
| 6     | 7     | 11    |
| 4     | 9     | 13    |

OBJECTIVE 2, PAGE 8

The data from Example 2 was from Table (a) in the video that presented a conceptual understanding of one-way ANOVA video, and the *F*-test statistic was 38.57.

For the data in Table (b) in the conceptual understanding of one-way ANOVA video, MST=45,MSE=24.1667, and the *F*-test statistic is 1.86.

The small *F*-test statistic for the data in Table (b) is evidence that the sample means for each treatment do not differ.

OBJECTIVE 2, PAGE 9
**Note:** The computations that lead to the *F*-test statistic are presented in Table 2, which is called an ANOVA table.

**Table 2**

| Source of Variation | Sum of Squares | Degrees of Freedom | Mean Squares | F-Test Statistic |
|---|---|---|---|---|
| Treatment | $SST$ | $k-1$ | $MST = \dfrac{SST}{k-1}$ | $F_0 = \dfrac{MST}{MSE}$ |
| Error | $SSE$ | $N-k$ | $MSE = \dfrac{SSE}{N-k}$ | |
| Total | $SST + SSE$ | $N-1$ | | |

OBJECTIVE 2, PAGE 11
12) State the decision rule for a one-way ANOVA test.

| Example 3 | *Performing One-Way ANOVA Using Technology* |
|---|---|

Prosthodentists specialize in the restoration of oral function, including the use of dental implants, veneers, dentures, and crowns. Since repairing chipped veneer is less costly and time consuming than complete restoration, a researcher wanted to compare the shear bond strength of different kits for repairs of chipped porcelain veneer in fixed prosthodontics. He randomly divided 20 porcelain specimens into four treatment groups. Group 1 specimens used the Cojet system, group 2 used the Silistor system, group 3 used the Cimara system, and group 4 specimens used the Ceramic Repair system. At the conclusion of the study, shear bond strength (in megapascals, MPa) was measured according to ISO 10477. The data in Table 3 are based on the results of the study. Do the data suggest that there is a difference in the mean shear bond strength among the four treatment groups at the $\alpha = 0.05$ level of significance?

**Table 3**

| Cojet | Silistor | Cimara | Ceramic Repair |
|---|---|---|---|
| 15.4 | 17.2 | 5.5 | 11.0 |
| 12.9 | 14.3 | 7.7 | 12.4 |
| 17.2 | 17.6 | 12.2 | 13.5 |
| 16.6 | 21.6 | 11.4 | 8.9 |
| 19.3 | 17.5 | 16.4 | 8.1 |

Data from P. Schmage et al. "Shear Bond Strengths of Five Intraoral Porcelain Repair Systems," Journal of Adhesion Science & Technology 21 (5–6):409–422, 2007

OBJECTIVE 2, PAGE 13

**Note:** Whenever performing analysis of variance, it is a good idea to present visual evidence that supports the conclusions of the test. Side-by-side boxplots are a great way to help visually reinforce the results of the ANOVA procedure.

OBJECTIVE 2, PAGE 15

**Note:** When we reject the null hypothesis of equal population means, as in Example 3, we know that at least one population mean differs from the others. However, we do not know which means differ.

Side-by-side boxplots can give us some idea, but we can more formally answer this question using Tukey's test, which will be discussed in the next section.

OBJECTIVE 2, PAGE 16

**Note:** Another way to verify the normality requirement in a one-way ANOVA test is through a residual plot. The residual of each value is the difference between the value and its sample's mean. Once the residual has been calculated for each value in each sample, create a normal probability plot for the residuals. If the residuals are found to be normally distributed, the normality requirement is satisfied.

**Example 4**     *Verifying the Normality Requirement by Analyzing Residuals*

Verify the normality requirement for the data analyzed in Example 3.

**Table 3**

| Cojet | Silistor | Cimara | Ceramic Repair |
|-------|----------|--------|----------------|
| 15.4  | 17.2     | 5.5    | 11.0           |
| 12.9  | 14.3     | 7.7    | 12.4           |
| 17.2  | 17.6     | 12.2   | 13.5           |
| 16.6  | 21.6     | 11.4   | 8.9            |
| 19.3  | 17.5     | 16.4   | 8.1            |

Data from P. Schmage et al. "Shear Bond Strengths of Five Intraoral Porcelain Repair Systems," *Journal of Adhesion Science & Technology* 21 (5–6):409–422, 2007

## Section 13.2
## Post-Hoc Tests on One-Way Analysis of Variance

**Objective**

❶ Perform Tukey's Test

---

Suppose the results of a one-way ANOVA show that at least one population mean is different from the others. To determine which means differ significantly, we make additional comparisons between means using procedures called multiple comparison methods.

---

### Objective 1: Perform Tukey's Test

OBJECTIVE 1, PAGE 1

1) What is compared in the Tukey test? What is the goal of the test?

2) State the formula for the standard error when using Tukey's test.

OBJECTIVE 1, PAGE 2

3) State the test statistic for Tukey's test when testing $H_0 : \mu_i = \mu_j$ *versus* $H_1 : \mu_i \neq \mu_j$.

OBJECTIVE 1, PAGE 3

**Note:** The critical value for Tukey's test using a familywise error rate α is given by

$$q_{\alpha, v, k}$$

where

$v$ is the degrees of freedom due to error, which is the total number of subjects sampled minus the number of means being compared, or $n - k$

$k$ is the total number of means being compared

We can determine the critical value from the Studentized range distribution by referring to Table X.

OBJECTIVE 1, PAGE 4

| Example 1 | *Finding Critical Values from the Studentized Range Distribution* |
|---|---|

Find the critical value from the Studentized range distribution with $v = 7$ degrees of freedom and $k = 3$ degrees of freedom, with a familywise error rate $\alpha = 0.05$.

OBJECTIVE 1, PAGE 6

4) After rejecting the null hypothesis $H_0 : \mu_1 = \mu_2 = \cdots = \mu_k$, what are the six steps when performing Tukey's test?

Step 1

Step 2

Step 3

Step 4

Step 5

Step 6

**Example 2**    *Performing Tukey's Test by Hand*

In Example 3 from Section 13.1, we rejected the null hypothesis $\mu_{\text{Cojet}} = \mu_{\text{Silistor}} = \mu_{\text{Cimara}} = \mu_{\text{Ceramic Repair}}$. Use Tukey's test to determine which pairwise means differ using a familywise error rate of $\alpha = 0.05$.

OBJECTIVE 1, PAGE 8

**Example 3**    *Performing Tukey's Test Using Technology*

In Example 3 from Section 13.1, we rejected the null hypothesis. Use StatCrunch to conduct Tukey's test to determine which pairwise means differ using a familywise error rate of $\alpha = 0.05$.

OBJECTIVE 1, PAGE 10

Sometimes the results of Tukey's test are ambiguous. Suppose the null hypothesis $H_0 : \mu_1 = \mu_2 = \mu_3 = \mu_4$ is rejected and the results of Tukey's test indicate the following:

$$\underline{\mu_1 \quad \mu_2} \quad \underline{\mu_3 \quad \mu_4}$$

We can conclude from this result that $\mu_1 = \mu_2 \neq \mu_4$, but we cannot tell how $\mu_3$ is related to $\mu_1$, $\mu_2$, or $\mu_4$. A solution to this problem is to increase the sample size so that the test is more powerful.

It can also happen that the one-way ANOVA rejects $H_0 : \mu_1 = \mu_2 = \cdots = \mu_k$ but the Tukey test does not detect any pairwise differences. The result occurs because one-way ANOVA is more powerful than Tukey's test. Again, the solution is to increase the sample size.

# Chapter 14 – Inference on the Least-Squares Regression Model

## OUTLINE

### Putting It Together

In Chapter 4, we learned methods for describing the relation between bivariate quantitative data. We also learned to perform diagnostic tests, such as determining whether a linear model is appropriate, identifying outliers, and identifying influential observations.

In this chapter, we begin by extending hypothesis testing and confidence intervals to least-squares regression models. In Section 14.1, we test whether a linear relation exists between two quantitative variables using methods based on those presented in Chapter 10. In Section 14.2, we construct confidence intervals about the predicted value of the response variable.

## Section 14.1
## Testing the Significance of the Least-Squares Regression Model

**Objectives**

❶ State the Requirements of the Least-Squares Regression Model

❷ Compute the Standard Error of the Estimate

❸ Verify That the Residuals Are Normally Distributed

❹ Conduct Inference on the Slope of the Least-Squares Regression Model

❺ Construct a Confidence Interval about the Slope of the Least-Squares Regression Model

INTRODUCTION, PAGE 2

| **Example 1** | ***A Review of Least-Squares Regression*** |

A family doctor is interested in examining the relationship between a patient's age and total cholesterol (in milligrams per deciliter). He randomly selects 14 of his female patients and obtains the data presented in Table 1. The data are based on results obtained from the National Center for Health Statistics. Draw a scatter diagram, compute the correlation coefficient, find the least-squares regression equation, and determine the coefficient of determination.

**Table 1**

| Age, $x$ | Total Cholesterol, $y$ | Age, $x$ | Total Cholesterol, $y$ |
|----------|------------------------|----------|------------------------|
| 25 | 180 | 42 | 183 |
| 25 | 195 | 48 | 204 |
| 28 | 186 | 51 | 221 |
| 32 | 180 | 51 | 243 |
| 32 | 210 | 58 | 208 |
| 32 | 197 | 62 | 228 |
| 38 | 239 | 65 | 269 |

---

### *Objective 1: State the Requirements of the Least-Squares Regression Model*

OBJECTIVE 1, PAGE 1

Because $b_0$ and $b_1$ are statistics, their values vary from sample to sample; so a sampling distribution is associated with each. We use this sampling distribution to perform inference on $b_0$ and $b_1$. For example, we might want to test whether $\beta_1$ is different from 0. If we have sufficient evidence to this effect, we conclude that there is a linear relation between the explanatory variable, $x$, and response variable, $y$.

1) What does the notation $\mu_{y|32}$ represent?

OBJECTIVE 1, PAGE 2

*Answer the following after watching the video.*

2) State the two requirements of the least-squares regression model.

3) What does the first requirement mean?

<u>OBJECTIVE 1, PAGE 2 (CONTINUED)</u>
4) How can you check if the first requirement is satisfied?

<u>OBJECTIVE 1, PAGE 3</u>
5) State the equation that the least-squares regression model is given by.

---

*Objective 2: Compute the Standard Error of the Estimate*

<u>OBJECTIVE 2, PAGE 1</u>
6) State the formula for the standard error of the estimate, $s_e$.

**Example 2**     *Determining the Standard Error*

Compute the standard error of the estimate for the age and total cholesterol data in Table 1.

**Table 1**

| Age, $x$ | Total Cholesterol, $y$ | Age, $x$ | Total Cholesterol, $y$ |
|----------|------------------------|----------|------------------------|
| 25 | 180 | 42 | 183 |
| 25 | 195 | 48 | 204 |
| 28 | 186 | 51 | 221 |
| 32 | 180 | 51 | 243 |
| 32 | 210 | 58 | 208 |
| 32 | 197 | 62 | 228 |
| 38 | 239 | 65 | 269 |

---

*Objective 3: Verify That the Residuals Are Normally Distributed*

7) How do we verify that the residuals are normally distributed?

---

**Example 3**     *Verify that Residuals Are Normally Distributed*

Verify that the residuals obtained in Example 1 are normally distributed.

| x | y | Residual | x | y | Residual |
|---|---|----------|---|---|----------|
| 25 | 180 | −6.33 | 42 | 183 | −27.12 |
| 25 | 195 | 8.67 | 48 | 204 | −14.51 |
| 28 | 186 | −4.53 | 51 | 221 | −1.71 |
| 32 | 180 | −16.12 | 51 | 243 | 20.29 |
| 32 | 210 | 13.88 | 58 | 208 | −24.50 |
| 32 | 197 | 0.88 | 62 | 228 | −10.10 |
| 38 | 239 | 34.48 | 65 | 269 | 26.70 |

---

*Objective 4: Conduct Inference on the Slope of the Least-Squares Regression Model*

8) If there is no linear relation between the response and explanatory variables, what will the slope of the true regression line be? Why?

OBJECTIVE 4, PAGE 1 (CONTINUED)

9) To conclude that a linear relation exists between two variables, what null and alternative hypotheses can be used?

OBJECTIVE 4, PAGE 2

10) State the test statistic for the slope in a least-squares regression model. What distribution does it follow?

OBJECTIVE 4, PAGE 3

11) What are the two conditions that must be satisfied before testing a hypothesis regarding the slope coefficient, $\beta_1$?

OBJECTIVE 4, PAGE 3 (CONTINUED)
12) State the five steps for testing a hypothesis regarding the slope coefficient, $\beta_1$.


Step 1




Step 2



Step 3 (By Hand)                              Step 3 (Using Technology)






Step 4



Step 5




OBJECTIVE 4, PAGE 4
**Note: Handling Departures from Normality**
Because these procedures are robust, minor departures from normality will not adversely affect the results of the test.

In fact, for large samples $(n \geq 30)$, inferential procedures regarding $b_1$ can be used even with significant departures from normality.

**Example 4**     *Testing for a Linear Relation*

Test whether a linear relation exists between age and total cholesterol at the $\alpha = 0.05$ level of significance using the data in Table 1 from Example 1.

**Table 1**

| Age, x | Total Cholesterol, y | Age, x | Total Cholesterol, y |
|--------|----------------------|--------|----------------------|
| 25 | 180 | 42 | 183 |
| 25 | 195 | 48 | 204 |
| 28 | 186 | 51 | 221 |
| 32 | 180 | 51 | 243 |
| 32 | 210 | 58 | 208 |
| 32 | 197 | 62 | 228 |
| 38 | 239 | 65 | 269 |

---

## Objective 5: Construct a Confidence Interval about the Slope of the Least-Squares Regression Model

OBJECTIVE 5, PAGE 1

13) State the formulas for the lower bound and upper bound of a $(1-\alpha)\cdot100\%$ confidence interval for the slope of the regression line, $\beta_1$.

OBJECTIVE 5, PAGE 2

### Example 5 *Constructing a Confidence Interval for the Slope of the Least-Squares Regression Line*

Determine a 95% confidence interval for the slope of the least-squares regression line for the data in Table 1 from Example 1.

**Table 1**

| Age, $x$ | Total Cholesterol, $y$ | Age, $x$ | Total Cholesterol, $y$ |
|---|---|---|---|
| 25 | 180 | 42 | 183 |
| 25 | 195 | 48 | 204 |
| 28 | 186 | 51 | 221 |
| 32 | 180 | 51 | 243 |
| 32 | 210 | 58 | 208 |
| 32 | 197 | 62 | 228 |
| 38 | 239 | 65 | 269 |

<u>OBJECTIVE 5, PAGE 4</u>

14) As the value of $\sum (x_i - \bar{x})^2$ gets larger, what is the impact on the value of $s_{b_1}$ ?

15) What does this result imply?

<u>OBJECTIVE 5, PAGE 5</u>

16) State the two basic reasons we are intentionally avoiding a discussion of inference on the correlation coefficient.

## Section 14.1A
### Using Randomization Techniques on the Slope of the Least-Squares Regression Line

**Objectives**

❶ Use Randomization to Test the Significance of the Slope of the Least-Squares Regression Model

---

*Objective 1: Use Randomization to Test the Significance of the Slope of the Least-Squares Regression Model*

<u>OBJECTIVE 1, PAGE 1</u>

1) State the correlation and the least-squares regression equation, where $x$ is the Zestimate and $y$ is the selling price.

<u>OBJECTIVE 1, PAGE 2</u>

We can use the randomization techniques of Chapter 11 to judge whether two quantitative variables are significantly associated.

However, rather than using the correlation coefficient to judge whether a linear relation exists between two quantitative variables, we are going to use the slope of the least-squares regression equation.

<u>OBJECTIVE 1, PAGE 3</u>

2) We would like to know if the slope of 1.3059 suggests that higher Zestimates correspond with a higher selling price, or is it possible the two variables are not positively associated and the slope is 0?
State the two possibilities in formulating the judgement.

3) State the null and alternative hypotheses associated with these two possibilities.

<u>OBJECTIVE 1, PAGE 3 (CONTINUED)</u>
4) State the slope and correlation coefficient for the randomly assigned data from StatCrunch.

5) What do the slope and correlation coefficient suggest about the association between the two variables?

<u>OBJECTIVE 1, PAGE 4</u>
We will use the Randomization Test for Slope applet in StatCrunch in order to determine how likely it is to observe a sample slope as extreme or more extreme than the one actually observed.

<u>OBJECTIVE 1, PAGE 5</u>
6) How many of the 5000 samples resulted in a slope of 1.3059 or higher? What is the *P*-value?

7) What conclusion can be drawn about the association between the two variables?

<u>OBJECTIVE 1, PAGE 6</u>
8) What is the shape of the distribution in the null model? Where is it centered?

OBJECTIVE 1, PAGE 7
9) State the five steps for testing hypotheses regarding the slope of the Least-Squares Regression Using Randomization.


Step 1



Step 2





Step 3




Step 4




Step 5

| Example 1 | Testing Hypotheses Regarding the Slope of the Least-Squares Regression Line Using Randomization |
|---|---|

A family doctor is interested in examining the relationship between a patient's age and total cholesterol (in milligrams per deciliter). He randomly selects 14 of his female patients and obtains the data presented in Table 2. The data are based on results obtained from the National Center for Health Statistics. Does the sample evidence suggest a linear relation exists between age and total cholesterol?

**Table 2**

| Age, $x$ | Total Cholesterol, $y$ | Age, $x$ | Total Cholesterol, $y$ |
|---|---|---|---|
| 25 | 180 | 42 | 183 |
| 25 | 195 | 48 | 204 |
| 28 | 186 | 51 | 221 |
| 32 | 180 | 51 | 243 |
| 32 | 210 | 58 | 208 |
| 32 | 197 | 62 | 228 |
| 38 | 239 | 65 | 269 |

Step 1

Step 2

Step 3

Step 4

Step 5

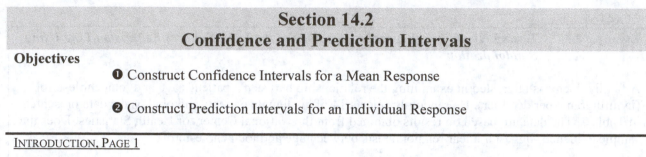

**Objectives**

❶ Construct Confidence Intervals for a Mean Response

❷ Construct Prediction Intervals for an Individual Response

INTRODUCTION, PAGE 1

1) State the two interpretations for the predicted value of total cholesterol $(\hat{y})$ for a given age $x$.

INTRODUCTION, PAGE 2

2) What is a confidence interval for a mean response?

3) What is a prediction interval for an individual response?

**Note:** Confidence intervals are used for a mean response and prediction intervals are used for an individual result.

## Objective 1: Construct Confidence Intervals for a Mean Response

4) State the formulas for the lower bound and upper bound for a $(1-\alpha)\cdot100\%$ confidence interval for $\hat{y}$, the mean response of $y$ for a specified value of $x$.

5) What are the required conditions for constructing a $(1-\alpha)\cdot100\%$ confidence interval for $\hat{y}$, the mean response of $y$ for a specified value of $x$?

**Example 1**    *Constructing a Confidence Interval for a Mean Response*

Construct a 95% confidence interval for the predicted mean total cholesterol of all 42-year-old females, using the data in Table 1.

**Table 1**

| Age, $x$ | Total Cholesterol, $y$ | Age, $x$ | Total Cholesterol, $y$ |
|----------|------------------------|----------|------------------------|
| 25 | 180 | 42 | 183 |
| 25 | 195 | 48 | 204 |
| 28 | 186 | 51 | 221 |
| 32 | 180 | 51 | 243 |
| 32 | 210 | 58 | 208 |
| 32 | 197 | 62 | 228 |
| 38 | 239 | 65 | 269 |

## *Objective 2: Construct Prediction Intervals for an Individual Response*

<u>OBJECTIVE 2, PAGE 1</u>

6) State the formulas for the lower bound and upper bound for a $(1-\alpha)\cdot 100\%$ prediction interval for $\hat{y}$, the individual response of $y$ for a specified value of $x$.

7) What are the required conditions for constructing a $(1-\alpha)\cdot 100\%$ confidence interval for $\hat{y}$, the individual response of $y$ for a specified value of $x$?

**Example 2**    *Constructing a Prediction Interval for an Individual Response*

Construct a 95% prediction interval for the predicted total cholesterol for a 42-year-old females, using the data in Table 1.

**Table 1**

| Age, $x$ | Total Cholesterol, $y$ | Age, $x$ | Total Cholesterol, $y$ |
|----------|------------------------|----------|------------------------|
| 25       | 180                    | 42       | 183                    |
| 25       | 195                    | 48       | 204                    |
| 28       | 186                    | 51       | 221                    |
| 32       | 180                    | 51       | 243                    |
| 32       | 210                    | 58       | 208                    |
| 32       | 197                    | 62       | 228                    |
| 38       | 239                    | 65       | 269                    |

8) Explain why the interval about the individual (prediction interval for an individual response) is wider than the interval about the mean (confidence interval for a mean response).